OCR (B) GCSE

Geography

Jane Ferretti
Brian Greasley

Philip Allan Updates, an imprint of Hodder Education, an Hachette UK company, Market Place, Deddington, Oxfordshire OX15 0SE

Orders
Bookpoint Ltd, 130 Milton Park, Abingdon, Oxfordshire OX14 4SB
tel: 01235 827827
fax: 01235 400401
e-mail: education@bookpoint.co.uk

Lines are open 9.00 a.m.–5.00 p.m., Monday to Saturday, with a 24-hour message answering service. You can also order through the Philip Allan Updates website: www.philipallan.co.uk

ISBN 978-1-4441-1047-0

First printed 2010
Impression number 5 4 3 2
Year 2014 2013 2012

Printed in Dubai

P02104 Hachette UK's policy is to use papers that are natural, renewable and recyclable products and made from wood grown in sustainable forests. The logging and manufacturing processes are expected to conform to the environmental regulations of the country of origin.

About this book

Revision is vital for success in your GCSE examination. No-one can remember what they learnt up to 2 years ago without a reminder. To be effective, revision must be planned. This book provides a carefully planned course of revision — here is how to use it.

The book	***The route to success***
Contents list	**Step 1** Check which topics you need to revise for your examination. Mark them clearly on the contents list and make sure you revise them.
Revision notes	**Step 2** Each section of the book gives you the facts you need to know for a topic. Read the notes carefully, and list the main points.
Key words	**Step 3** Key words are given at the end of each information page and highlighted in the text. Learn them and their meanings. They must be used correctly in the examination.
Case study	**Step 4** Each section has a case study. Learn one for each topic. You could do this by listing the main headings with key facts beneath them on a set of revision cards. If your teacher has taught different case studies, use the one you find easiest to remember.
Test yourself	**Step 5** A set of brief questions is given at the end of each section. Answer these to test how much you know. If you get one wrong, revise it again. You can try the questions before you start the topic to check what you know.
Examination questions	**Step 6** Examples of questions are given for you to practise. Notice the higher-tier question at the end for those entered for the higher tier. The more questions you practise, the better you will become at answering them.
Exam tips	**Step 7** The exam tips offer advice for achieving success. Read them and act on the advice when you answer the question.
Key word index	**Step 8** On pp. 137–139 there is a list of all the key words and the pages on which they appear. Use this index to check whether you know all the key words. This will help you to decide what you need to look at again.

Command words

All examination questions include **command** or **action** words. These tell you what the examiner wants you to do. Here are some of the most common ones.

- **List** — usually wants you to provide a list of facts.
- **Describe** — requires more than a list. For example, you are expected to write a description of the pattern of population in Wales, but not to give any explanation for it.
- **Explain**, **give reasons for** or **account for** — here the examiner is expecting you to show understanding by giving reasons, and to do more than describe, for example, the pattern of population in Wales.
- **Suggest** — the examiner is looking for sensible explanations, using your geographical knowledge, for something to which you cannot know the actual answer — for example, 'Suggest reasons for the location of the factory in photograph A'.
- **Compare** — the best candidate will not write two separate accounts of the factors to be compared, but will pick several points and compare them one at a time. Useful phrases to use are 'whereas', 'on the other hand', 'compared to'.

Do you know?

- The exam board setting your paper?
- What level or tier you will be sitting?
- How many papers you will be taking?
- The date, time and place of each paper?
- How long each paper will be?
- What the subject of each paper will be?
- What the paper will look like? Do you write your answer on the paper or in a separate booklet?
- How many questions you should answer?
- Whether there is a choice of questions?
- Whether any part of the paper is compulsory?

If you don't know the answer to any of these questions as the exam approaches — ask your teacher!

Revision rules

- Start early.
- Plan your time by making a timetable.
- Be realistic — don't try to do too much each night.
- Find somewhere quiet to work.
- Revise thoroughly — reading is not enough.
- Summarise your notes, make headings for each topic, and list the case study examples.
- Ask someone to test you.
- Try to answer some questions from old papers. Your teacher will help you.

If there is anything you don't understand — ask your teacher.

Be prepared

The night before the exam

- Complete your final revision.
- Check the time and place of your examination.
- Get ready your pens, pencil, coloured pencils, ruler and calculator (if you are allowed to use one).
- Go to bed early and set the alarm clock!

On the examination day

- Don't rush.
- Double check the time and place of your exam and your equipment.
- Arrive early.
- Keep calm — breathe deeply.
- Be positive.

Examination tips

- Keep calm and concentrate.
- Read the paper through before you start to write.
- If you have a choice, decide which questions you are going to answer.
- Make sure you can do all parts of the questions you choose, including the final sections.
- Complete all the questions.
- Don't spend too long on one question at the expense of the others.
- Stick to the point and answer questions fully.
- Use all your time.
- Check your answers.
- Do your best.

Theme 1

Rivers and coasts

The hydrological cycle

The hydrological cycle or the water cycle is the continuous movement of water between the land, the sea and the atmosphere.

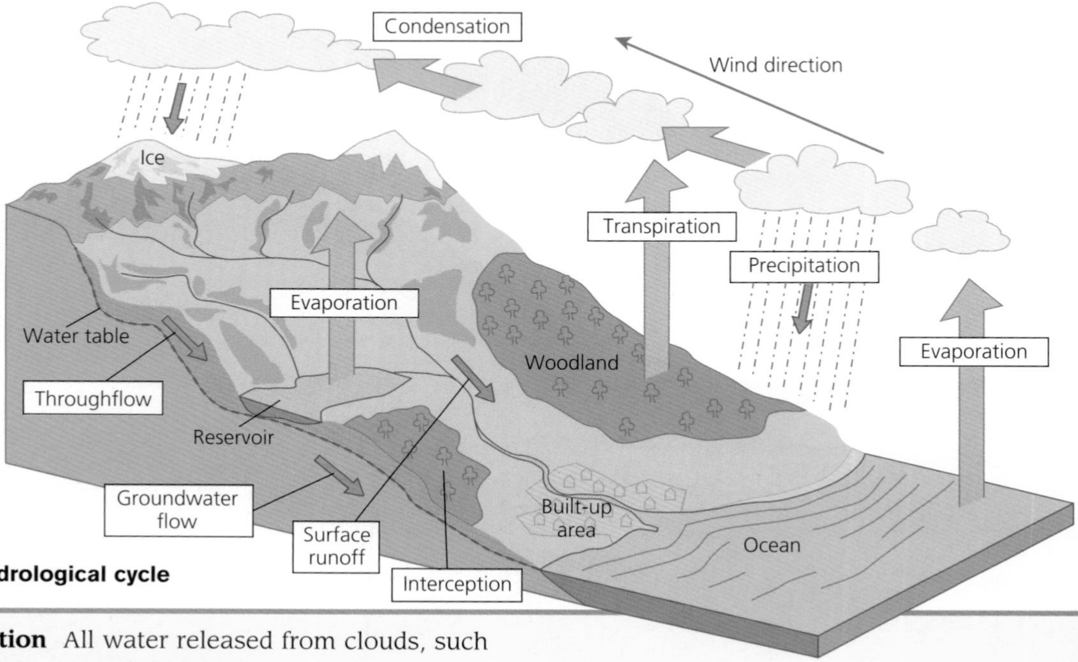

The hydrological cycle

Precipitation All water released from clouds, such as rain, snow, hail, sleet and fog.

Surface runoff Water flowing across the surface. The water may be in a channel, such as a river or stream, or it may be overland flow when it makes its way across a field or down a roadway.

Interception When water collects on objects such as leaves or flat roofs.

Infiltration When water soaks into soil.

Throughflow When water soaks into soil and seeps through it towards a river or the sea.

Percolation The downward movement of water through soil into rocks.

Groundwater flow The movement of water below the water table. Water which is stored underground is called groundwater.

Evaporation When water which is heated by the Sun becomes vapour and rises into the atmosphere. This may take place over land or sea.

Transpiration All plants lose water through their leaves. Transpiration is when this water returns to the atmosphere where it evaporates.

Evapotranspiration The term for the processes of both evaporation and transpiration.

Condensation When water vapour is cooled and turns into water droplets to form clouds.

Water table The upper level of saturated ground. It is not a 'table'. It is not even flat. The level is closer to the surface in winter when there is plenty of rain.

The drainage basin

The **drainage basin** is the area of land drained by a river and its **tributaries**. Large rivers such as the Nile or Mississippi have vast drainage basins. The boundary of a drainage basin is called the **watershed**. This is usually a range of hills or mountains. Rain falling beyond the watershed will flow into another river and is part of another drainage basin.

A drainage basin can also be represented in a systems diagram such as the one below. This shows how water is held in stores and how it is transferred between stores.

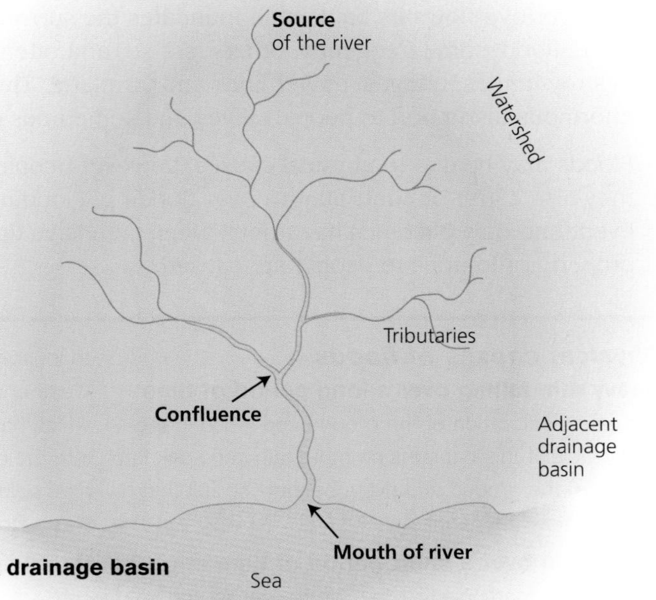

A drainage basin

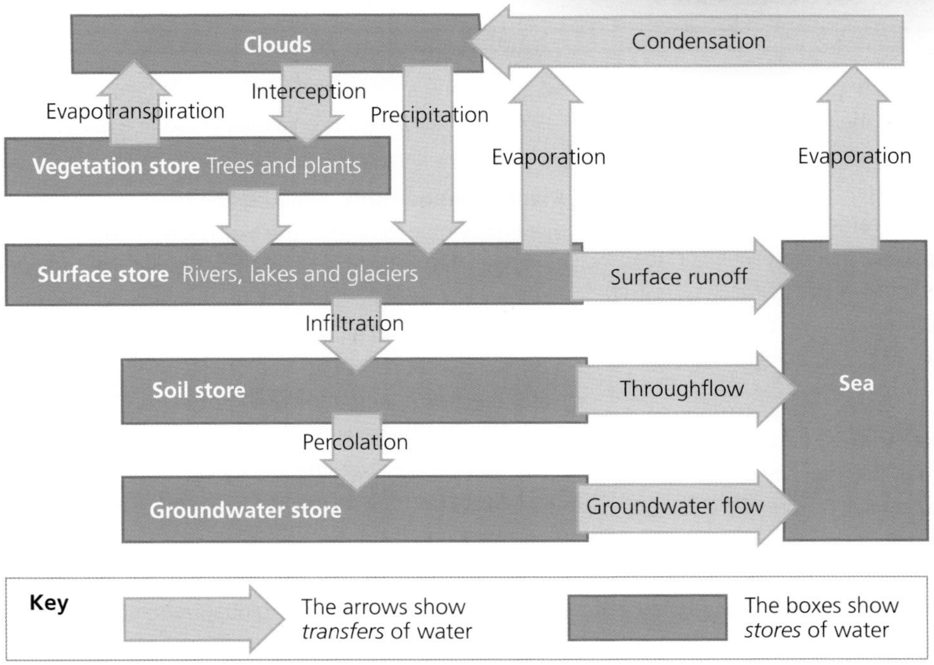

The drainage basin as a system

Key words

- precipitation
- surface runoff
- interception
- infiltration
- throughflow
- percolation
- groundwater flow
- evaporation
- transpiration
- evapotranspiration
- condensation
- water table
- drainage basin
- tributary
- watershed
- source
- confluence
- mouth of river

Causes of flooding

If a river overflows its banks and inundates the surrounding land it is called a flood. Many rivers flood, some frequently, others less so, and often with little warning. Severe floods can destroy homes, bridges, power lines and farmland. They can cause enormous disruption to people's lives and sometimes even death.

Floods may be due to physical causes, however people's activities may also cause or contribute to river flooding. Flooding is a normal event and only becomes hazardous when extensive damage to property or injuries to people are caused.

Key words

- flash flood
- urbanisation
- deforestation
- river management

Physical causes of floods

Heavy rain falling over a long period of time can cause saturation of the ground and a rise in the water table. If this happens no more rain can soak into (infiltrate) the ground and rivers become so full that they overflow their banks.

Heavy rain over a short period of time can cause floods particularly if the rain is intense and falls on hard ground which does not allow water to infiltrate. This can happen if there has been hot weather in the days or weeks before the rainstorm which has baked the ground. Rainfall cannot soak into the ground so it flows quickly into rivers which rise rapidly and overflow their banks resulting in a **flash flood**.

Rapid snow melt in late winter and spring may cause flooding as the amount of water flowing into the river is simply too much for the river to hold.

People's activities which can cause flooding

Urbanisation In recent years the demand for houses has led to more building on land beside rivers. This land may be the river's natural floodplain, meaning it is at risk of flooding. Building leads to land being covered in tarmac or concrete which are impermeable surfaces, this increases surface runoff so more water flows into the river. Drains in built-up areas channel surface water directly into the river so that lag time is reduced and the river rises quickly, making it more likely to flood.

Deforestation If trees are removed there is little interception and little to break the intensity of rainfall hitting the ground and this may cause soil erosion. Eroded soil is washed into river channels and may be deposited in the channel in lowland areas, reducing the capacity of the river and making flooding more likely. Deforestation has occurred in many less economically developed countries (LEDCs) in recent decades, for example in tropical rainforest areas and in the Himalayas in Nepal. Trees are felled because timber is a valuable resource either for sale or for use as firewood.

River management Although the main aim of river management is to reduce the likelihood of flooding, in some situations it can lead to an increased risk of flooding. For example, straightening a river channel and lining it with concrete can mean an area further downstream which is not managed becomes at greater risk of flooding because water reaches it more rapidly and overwhelms the channel. Artificial embankments (levées) built alongside a river to keep water in the river channel can be breached, and this can cause severe flooding, as happened in New Orleans after Hurricane Katrina in 2005.

It is likely that flooding will increase in the future as a result of climate change caused by global warming. Changing patterns of rainfall and the increased number of storms will increase flood risk.

The storm hydrograph

A storm **hydrograph** shows how river discharge changes over a short period after a rainstorm. **Discharge** is the amount of water in a river. Storm hydrographs show how a river responds to heavy rainfall so they can be used to predict floods and help planners to decide about flood prevention strategies.

A hydrograph is divided into two parts, **base flow**, which is water entering the stream from ground storage, and **storm flow**, which is the water from the recent rainstorm. Base flow remains fairly constant but storm flow changes, so it is this part of the graph which shows how the river level changes as a result of a rainstorm.

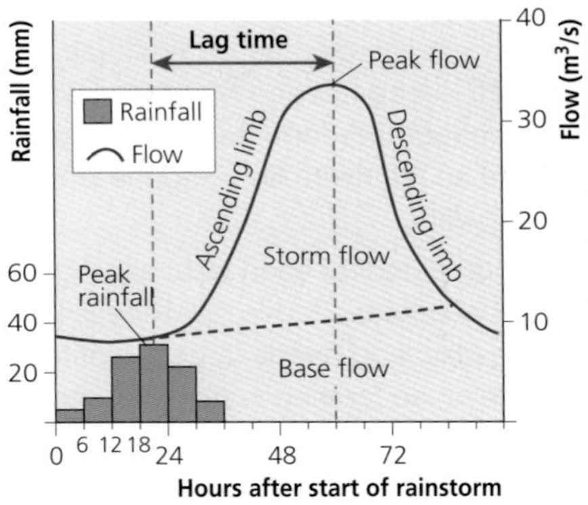

A storm hydrograph

The **ascending limb** shows how river discharge increases and then reaches **peak discharge** (the highest amount of water in the river). The time between peak rainfall and peak discharge is known as the **lag time**. The **descending limb** shows how river levels return to normal.

The shape of a hydrograph can vary. The steeper the ascending limb and the shorter the lag time, the more likely the river is to flood because the amount of water in the river is rising quickly. A hydrograph with a longer lag time shows a river that is less likely to flood.

Key words

- hydrograph
- discharge
- base flow
- storm flow
- ascending limb
- peak discharge
- lag time
- descending limb

Hydrograph in a woodland

This hydrograph has a gentle ascending limb because:

- trees and leaves intercept rain so it is less likely to reach the river
- the ground is soft and water easily infiltrates so reducing surface runoff
- the water table is low as trees take up soil moisture

Flooding is unlikely in wooded areas.

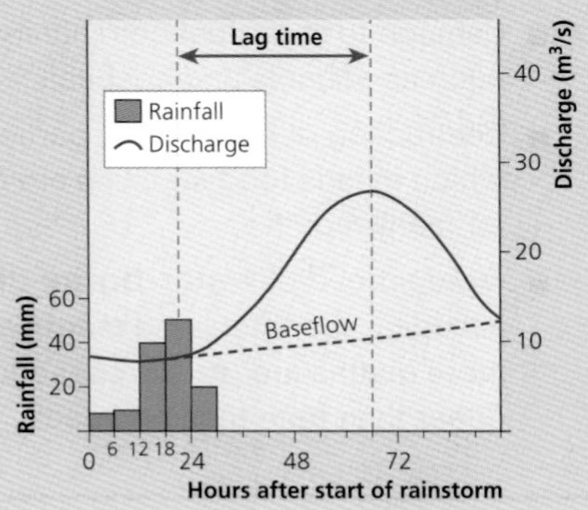

Hydrograph in an urban area

This hydrograph has a steep ascending limb because:

- the ground surface is covered by tarmac or concrete and is impermeable so there is little infiltration and a lot of surface runoff
- drains carry rainwater to the river very quickly

Flooding is more likely in an urban area.

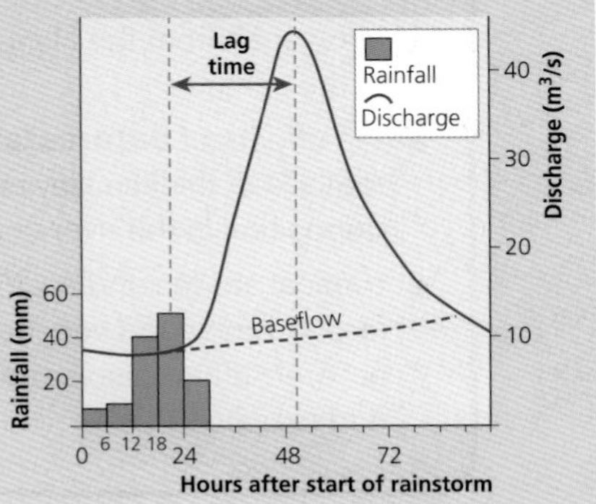

Case Study: *causes, effects and management of river flooding in an LEDC*

Ganges–Brahmaputra river basin

The River Ganges rises in the Himalayas and flows through northern India into Bangladesh where it joins the Brahmaputra and Meghna rivers which together form the vast Ganges Delta as the rivers flow into the Bay of Bengal.

Flooding occurs each year, leaving behind valuable fertile silt. Rice is the main crop and needs large quantities of water, so farming is planned around the floods. However, in recent years floods have become bigger and lasted longer, resulting in huge problems for people who live close to the rivers, particularly in Bangladesh.

Bangladesh lies almost entirely on the Ganges Delta and most of the country is flat and low-lying — almost all of it is lower than 12 metres above sea level. It is at risk from rising sea levels and floods.

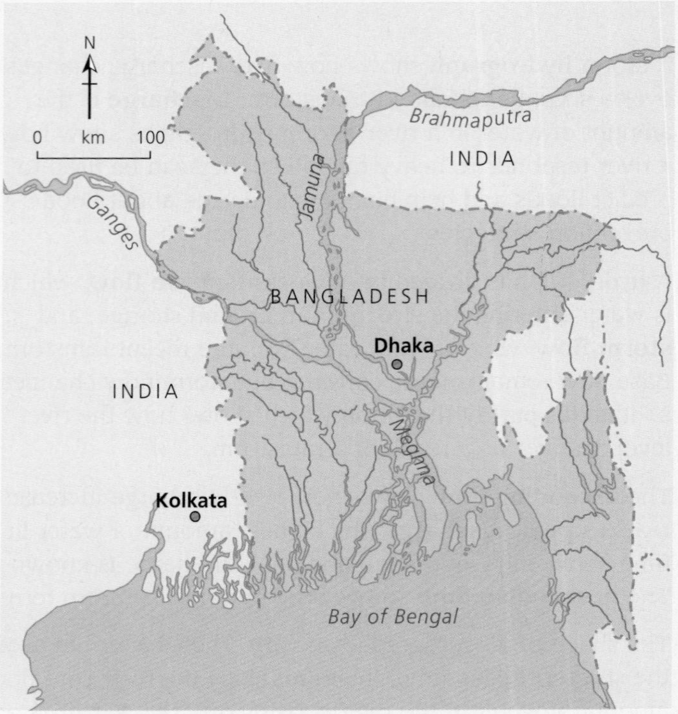

Location of Rivers Ganges, Brahmaputra and Meghna

In 2004, 36 million people were made homeless and 800 people died following serious flooding and three years later, in 2007, there were 500 deaths as a result of flooding.

Flooding is caused by a combination of:

- the **monsoon** climate which brings heavy rain between June and October
- snow melt from the Himalayas particularly in the summer
- the effects of **deforestation** in the foothills of the Himalayas causing deposition in the river channel downstream
- tropical storms and cyclones which cause strong winds and very heavy rain that can severely affect discharge in the three rivers and cause floods

The main effects of flooding in Bangladesh are:

- injuries and loss of life
- loss of, and damage to, houses, causing many people to become homeless
- major disruption to transport as roads and railways are flooded
- damage to important community buildings such as schools and hospitals
- destruction of food crops, particularly rice, which means food supplies are disrupted for many months
- rapid spread of **water-borne diseases** including dysentery and diarrhoea. Many more deaths are usually caused by disease rather than from the flood itself

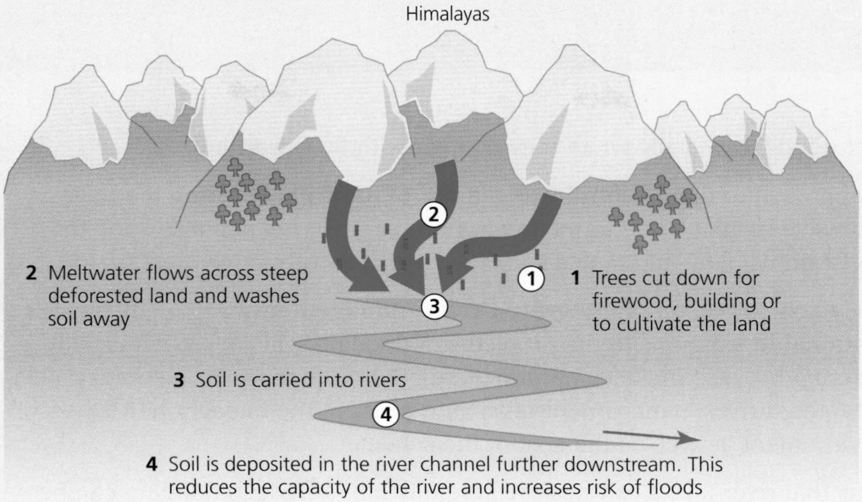
Himalayas
2 Meltwater flows across steep deforested land and washes soil away
1 Trees cut down for firewood, building or to cultivate the land
3 Soil is carried into rivers
4 Soil is deposited in the river channel further downstream. This reduces the capacity of the river and increases risk of floods

The effects of deforestation in the Himalayas

Bangladesh is one of the countries most vulnerable to **climate change**. It has been estimated that there could be 25 million climate refugees in the future. Climate change also increases the frequency and severity of tropical storms.

Managing flooding in the Ganges–Brahmaputra river basin

Bangladesh is a poor country and often has to rely on aid from other countries and from non-governmental organisations such as Oxfam and Save the Children after serious flooding. Short-term emergency aid brings medical and food supplies to the flooded area and over the longer term aid is used to help the country rebuild its infrastructure.

In the past ten years earth **embankments** up to 7 metres high have been constructed along 6,000 kilometres of the river. However, these have not been successful because:

- the embankments are easily breached
- they tend to create a false sense of security for people living nearby
- they make it difficult for floodwater to flow back into the river and therefore prolong floods

Now it has been decided that management strategies should aim to lessen the damage rather than control the floods. This is being done by:

- having better **flood warning** systems so people can protect themselves
- building **flood shelters** on stilts in areas prone to flooding to provide somewhere safe for people to go to during floods
- planting vegetation along the banks of the river to increase interception when it rains and reduce the amount of groundwater
- dredging the river channels and opening up abandoned channels to speed the flow of water away from the area

In the long term it is important for Bangladesh to work with India and Nepal to develop an **integrated plan** to tackle the problem of flooding in the Ganges–Brahmaputra basin.

Key words

monsoon	embankments
deforestation	flood warning
water-borne disease	flood shelter
climate change	integrated plan

Flood control

It is difficult to prevent floods, especially when they happen with little warning.

In the past the usual approach to flood control was the use of **hard engineering** methods. These are usually expensive to install and to maintain. Hard engineering can also spoil the environment and impact on wildlife, and may also make rivers less attractive.

Today, the approach is likely to be one of **integrated river management**, which emphasises the importance of **sustainable** solutions to the problems of flooding. This approach usually has some hard engineering but also includes natural flood control methods (sometimes called **soft engineering**). Integrated river management takes into account the impact on the lives of people and on the environment as well as the cost of the scheme.

New technology means that river levels can now be carefully monitored, so flooding can be predicted and warnings given to people living in areas likely to be flooded. In Britain, this is the job of the Environment Agency. The risk of flooding can also be reduced by:

- Planting trees (**afforestation**) in the upper catchment. However, this changes the appearance and ecosystem of the area.
- Building a dam and reservoir. This is a very effective way of controlling flooding, and the new lake can be used for recreation and even for generating hydroelectricity. However, it is expensive. It leads to loss of land and sometimes displaces a lot of people, as in the Three Gorges Dam project in China.

Hard engineering

- Building embankments, also called 'dykes' or 'levées'. These raised banks increase the capacity of the channel and keep floodwaters in the river.
- **Channelisation**: straightening and deepening the river. Water flows more quickly along a straight channel, reducing the risk of flooding. However, areas further downstream may be at greater risk because the floodwaters reach them more quickly. The river will revert to its natural course unless the new channel is reinforced continually.
- Constructing dams and reservoirs in the upper sections of the river to trap water which can be released slowly, reducing the risk of flooding.

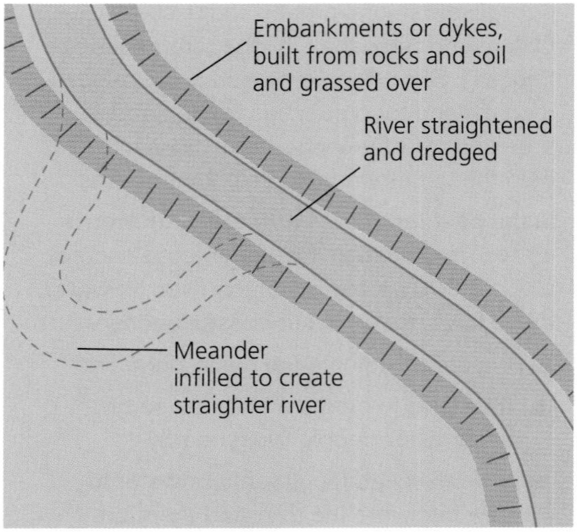

Hard engineering solutions

Natural flood control (soft engineering)

- Setting embankments back from the channel and avoiding building on floodplains to allow flooding around the river.
- Planting water-loving plants, such as willow and alder. This helps to lower the water table and increases the amount of wildlife.
- Digging a flood relief channel so the river can cope with increased discharge when necessary. In normal conditions, the meandering river course is maintained.
- Providing general maintenance for rivers to keep the water flowing, for example cleaning rubbish and debris from the channel, cutting back trees and repairing river banks.
- Dredging rivers to remove silt, sand and gravel to enlarge the channel and increase the amount of water it can hold.

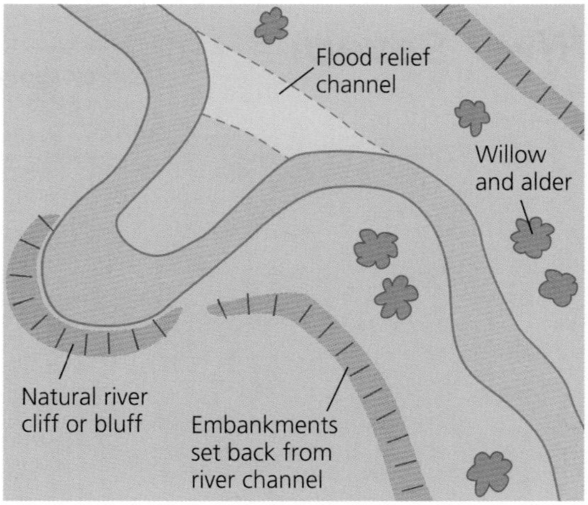

A natural approach to flood control

The natural approach enhances the environment and can provide access for people to enjoy the river and its surroundings. Natural solutions to flood control are less expensive to implement and maintain than are hard engineering solutions but, nevertheless, the Environment Agency does not have enough funding to do as much maintenance work as it would like.

Natural or soft engineering approaches are usually a more sustainable way to protect areas from floods.

Flood warnings

More than 5 million people in England and Wales live in properties that are at risk of flooding. One of the jobs of the Environment Agency is to provide information for these households and to issue warnings. It uses the latest technology to monitor rainfall and river levels 24 hours a day and uses this information to decide if **flood warnings** should be issued. The Environment Agency provides guidance on its web site and through leaflets. This guidance includes the following advice:

- Check the details of your insurance policy and add cover if necessary.
- Make an emergency kit and keep your mobile phone charged.
- Prepare a supply of sandbags if floods are forecast.
- Move all electrical goods and valuables to high shelves or upstairs if floods are imminent.

Key words

hard engineering
integrated river management
sustainable
soft engineering
afforestation
channelisation
flood warning

Case study: *causes, effects and management of river flooding in a MEDC*

River Severn

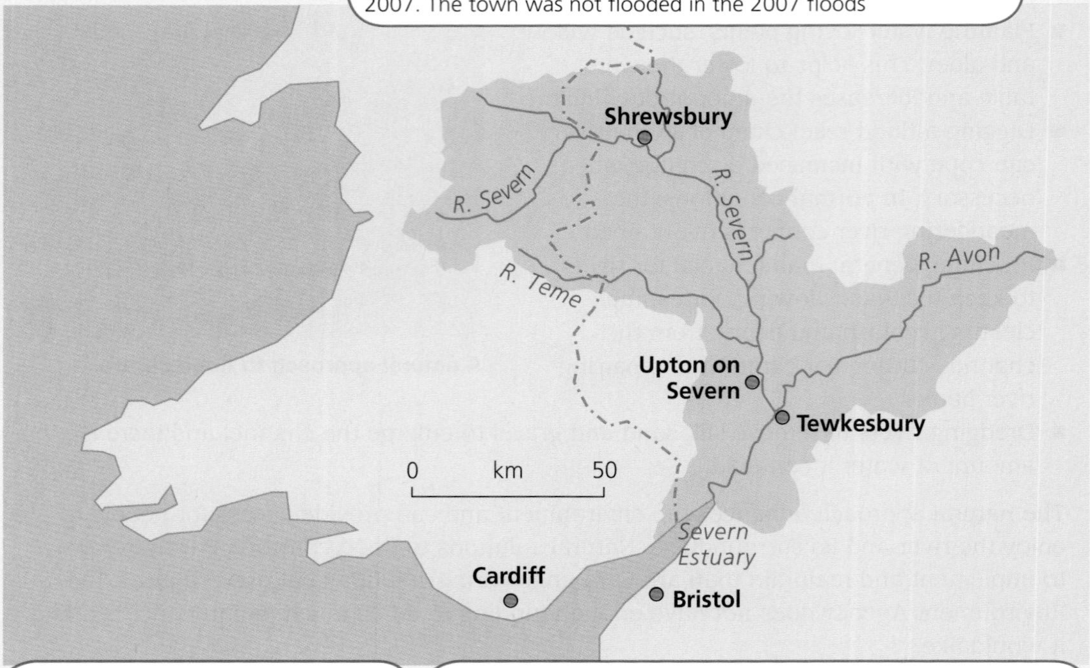

River Severn and its main tributaries

The River Severn is 220 miles long and is Britain's longest river. It rises in the Cambrian Mountains of mid Wales and flows into the Severn Estuary and the Bristol Channel. It has several major tributaries, including the River Avon and River Teme, and a number of large towns and cities lie along its route.

The River Severn floods regularly for a number of reasons:

- The Cambrian Mountains where the river rises is an area of high rainfall.
- It has a large catchment area and several large tributaries channel water into the River Severn.
- It flows across impermeable rock for much of its length so water cannot infiltrate and runs rapidly into the river.
- In recent years towns by the river have grown and floodplains have been used for new buildings, increasing surface runoff.
- Flood defences in some places (such as Shrewsbury) increase the flood risk in other areas further downstream.

July 2007 floods

Causes: During the summer of 2007 many parts of Britain were flooded. The floods along the River Severn were caused by heavy rain over a prolonged period resulting in the surrounding land becoming saturated and causing increased surface runoff. The river channel could not contain all the water and it overtopped its banks. However, flooding was probably made worse by flood defence schemes that have been erected in some areas and by building on the floodplain, which would otherwise provide a natural overspill area for water from the river.

Effects: The floods caused enormous damage to properties along the Severn, particularly in Upton, Tewkesbury and Gloucester. Three people died and thousands had to leave their homes. Many houses and businesses were badly damaged by rising floodwaters and, in places, sewerage and drainage systems collapsed. Roads and railway lines were closed and stranded people had to be rescued by the emergency services. A water treatment centre at Tewkesbury and an electricity substation at Gloucester were closed, leaving people without water or electricity for several days.

In the longer term, some people were still unable to return to their homes more than a year later. Farmers who lost their crops had no income and nothing to invest in their farms for the future, and some shops and businesses lost so much stock they had to close. Although insurance claims helped many to recover from the floods not everybody was adequately insured against all the damage.

Management

The Environment Agency is responsible for the UK's flood defences, although its budget is controlled by the government. The River Severn has extensive flood protection schemes in place and there are plans to improve these in the future.

In **Shrewsbury** the river has been straightened to allow the water to flow through the town more quickly. Parts of the town are protected by both embankments and flood walls and when floods are predicted removable barriers can be put in place. More flood walls and removable barriers are planned.

Upton on Severn is currently protected by temporary flood barriers that are put in place when river levels start to rise. However, in July 2007 the barriers did not arrive in time because they were delayed by hold-ups on the motorway, allowing floodwaters to rise and inundate the town. A £3.6 million scheme has been proposed which includes building a wall running 400 metres alongside the river. Although residents know that a permanent solution to the flooding problem must be found this plan is controversial, as many people do not want a permanent structure that will take away the riverside views which many visitors come to enjoy.

Tewkesbury has no engineered flood defences and relies on flood warnings provided by the Environment Agency. Since the devastation of 2007 local people have asked for proper flood defences to be put in place but at the moment there are no plans for this. Between now and 2026, just under 15,000 new homes are scheduled to be built in and around Tewkesbury. If these are built on the floodplain the flood risk for the town is likely to increase.

River processes

Like glaciers and the sea, rivers are important agents of **erosion**, **transport** and **deposition**.

Erosion

Rivers erode in four main ways:

- **Hydraulic action** The power of running water undercuts the banks and erodes the bed.
- **Abrasion** or **corrasion** Rocks and pebbles carried by the river crash against the sides and bed of the river, scraping away material.
- **Attrition** Rocks and pebbles carried by the river bang into each other and break up into smaller pieces.
- **Solution** or **corrosion** Some minerals from rocks are dissolved and carried away by the river, e.g. calcium carbonate.

The upper course of the river

Rivers usually rise in hills or mountains. The steep gradient of the river means that erosion, particularly **downward erosion**, is the dominant process. The river flows quickly towards the sea, carrying large amounts of **sediment** downstream. The river's course is not straight. It flows around **interlocking spurs** of higher land. There is a lot of **bedload** and many large rocks are angular in shape. Waterfalls and rapids are a common feature in this part of the river.

As waterfalls erode, they move slowly upstream, leaving a steep gorge on the lower side of the falls. An example is High Force on the River Tees, see p. 21.

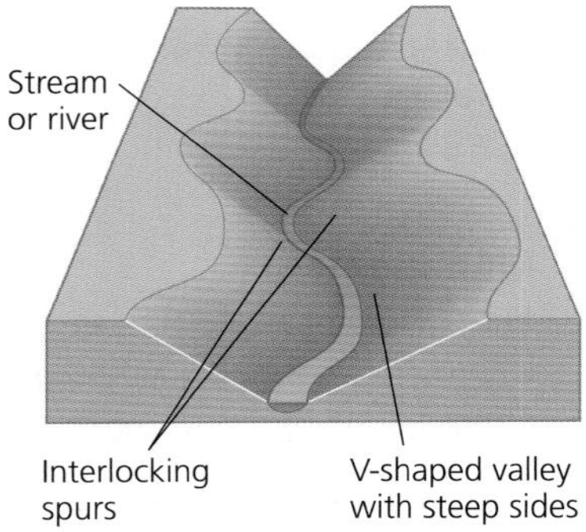

Interlocking spurs

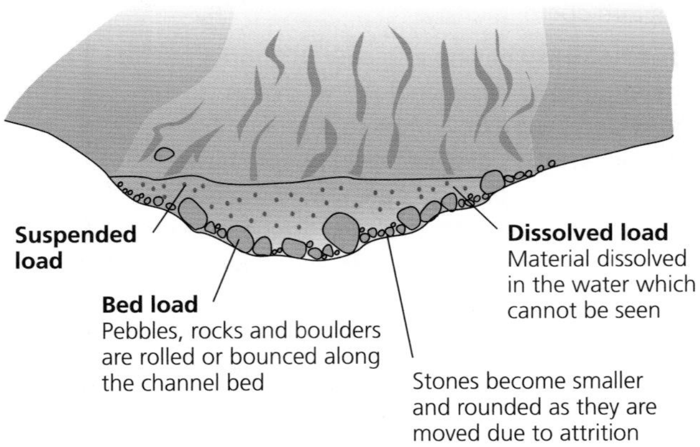

Cross-section through a river

Transport

Rivers transport an enormous load. This material can be carried in four ways:

- **Traction** Boulders and rocks are dragged or rolled along the river bed.
- **Saltation** Smaller-sized particles, such as pebbles or sand, are bounced along the bed.
- **Suspension** Fine particles of silt or clay are held in the water.
- **Solution** Minerals are dissolved in the river water.

The middle and lower course of the river

As the river flows through lower-lying land it flows faster and with less turbulence. Compared with the upper course, this section has a gentler gradient and higher discharge. The bedload is smaller and rounder. **Lateral** (sideways) **erosion** is more important than downward erosion, but deposition also occurs. Features such as meanders, oxbow lakes, floodplains and levées develop.

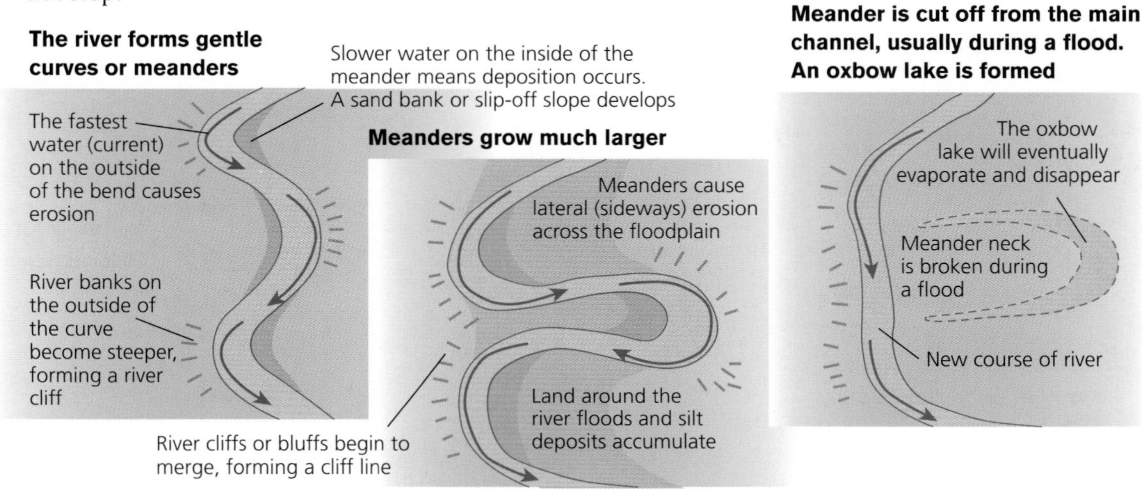
The river forms gentle curves or meanders
The fastest water (current) on the outside of the bend causes erosion
Slower water on the inside of the meander means deposition occurs. A sand bank or slip-off slope develops
River banks on the outside of the curve become steeper, forming a river cliff
Meanders grow much larger
Meanders cause lateral (sideways) erosion across the floodplain
Land around the river floods and silt deposits accumulate
River cliffs or bluffs begin to merge, forming a cliff line
Meander is cut off from the main channel, usually during a flood. An oxbow lake is formed
The oxbow lake will eventually evaporate and disappear
Meander neck is broken during a flood
New course of river

Deposition

A river drops its load when it no longer has enough energy to carry it. The larger, heavier material is deposited first, and this is seen higher up the river. Pebbles, gravel, sand and silt are deposited in the middle and lower course. The dissolved load is not deposited but is carried out to sea.

River floodplain

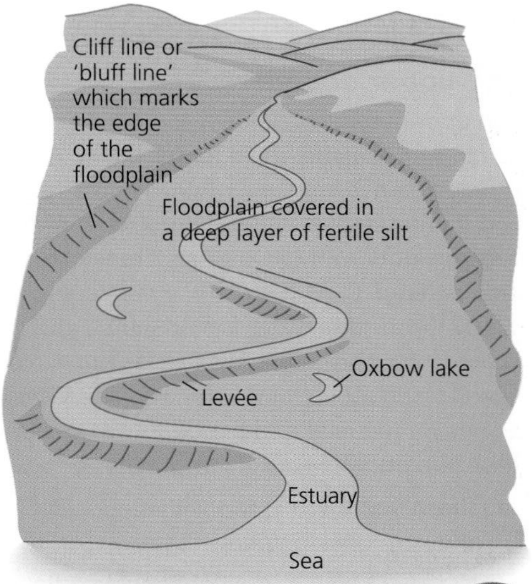
Cliff line or 'bluff line' which marks the edge of the floodplain
Floodplain covered in a deep layer of fertile silt
Oxbow lake
Levée
Estuary
Sea

The mouth of the river

As the river enters the sea, it tends to form a wide funnel shape, called an **estuary**. This part of the river is tidal and has a mixture of fresh water and salt water. Deposition of fine silt and clay around the estuary can form extensive mudflats and salt marshes.

Key words

erosion	abrasion/corrasion	sediment	saltation
transport	attrition	interlocking spurs	suspension
deposition	solution/corrosion	bedload	lateral erosion
hydraulic action	downward erosion	traction	estuary

Case study: *a river and its landforms*

The River Tees

The River Tees rises on Cross Fell in the Pennines about 890 metres above sea level and flows for 85 miles to its mouth in the North Sea between Redcar and Hartlepool.

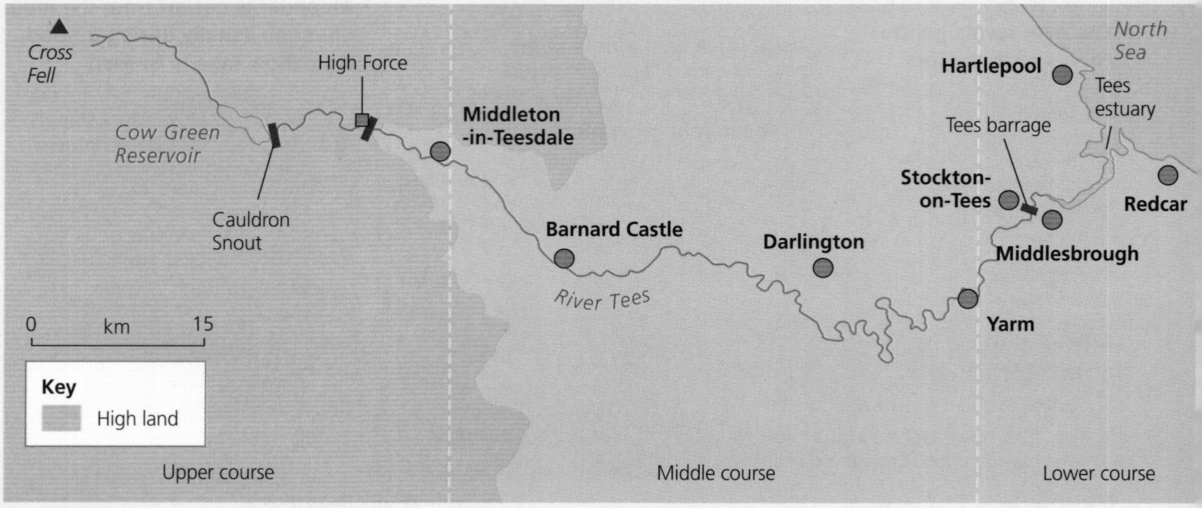

The River Tees

The upper course

The upper course is known as Teesdale and it is mostly moorland, used for hill sheep farming. Precipitation is high, over 2,000 mm a year, and the bedrock is impermeable, so the River Tees collects a lot of water as it runs steeply downstream. Cow Green reservoir, completed in 1971, provides water for industries and towns in the lower part of the Tees valley but is also used to control flooding. The Pennine Way runs alongside the river and the beautiful scenery in Teesdale attracts walkers and visitors.

- The valley is V-shaped with steep valley sides and interlocking spurs.
- The river has a narrow channel and flows fast because of the steep gradient.
- There are a number of **rapids** and **waterfalls**, particularly Cauldron Snout and High Force, caused by the river flowing over harder, impermeable rock.
- There are many small tributaries.
- The bedload is mainly large and angular boulders.
- Cauldron Snout is a series of cataracts where the river rushes over a rocky 'stairway', dropping about 180 metres in height.
- High Force is a spectacular waterfall which drops 20 metres vertically. It has formed where the Tees crosses the igneous rocks of the Whin Sill. There are two layers of rock at High Force; the upper layer is whinstone, which is a hard rock that erodes very slowly, and the lower layer is carboniferous limestone, which is more easily eroded by the river. The waterfall is slowly moving upstream, leaving a narrow, deep gorge downstream which is about 700 metres long.

The middle course

From Middleton-in-Teesdale, past Barnard Castle to Yarm the river is surrounded by lower undulating land which is used for arable farming. In this part of the river valley:

- the valley sides slope more gently and the valley floor is wider
- lateral erosion is more important than downward erosion
- the river channel is wider and the gradient is less steep
- the river meanders forming river cliffs or bluffs
- the bedload is smaller and more rounded

The lower course

Downstream from Yarm the river flows past Stockton-on-Tees and Middlesbrough before forming an **estuary** as it reaches the sea. The land on either side of the river is flatter and the area has become important for industry, particularly chemicals and steel making.

- The river channel is deep and narrow.
- Lateral erosion predominates.
- The load is mainly sand, gravel and alluvium.
- Along both sides of the valley floor is a wide **floodplain** which is covered by alluvium deposited by past floods.
- There are **meanders** and **oxbow lakes**.
- The river is tidal and forms an estuary.

In the early 1800s the river was diverted through two 'cuts' between Stockton and Middlesbrough. This bypassed the meanders and made the river straighter so that navigation was quicker and cheaper. Since then, the river has continued to be dredged and to have alterations to make the channel deeper and narrower, and it has been reduced from about 300 metres wide to 200 metres or less. Teesport is built on reclaimed land on the south side of the Tees estuary.

At Stockton-on-Tees a barrage was built which opened in 1995. The barrage has four huge flood gates which prevent the tidal water getting upstream, keeping the water level constant. This part of the river is now much cleaner and more attractive, and the surrounding land has been landscaped and water sports facilities have been developed. The barrage and land reclamation has helped the regeneration of Teesside.

High Force on the River Tees

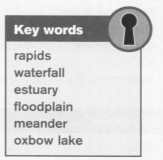

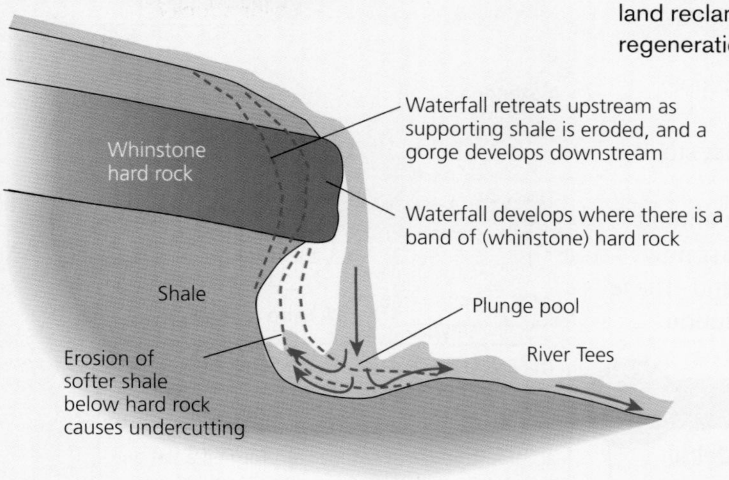
Whinstone hard rock
Waterfall retreats upstream as supporting shale is eroded, and a gorge develops downstream
Waterfall develops where there is a band of (whinstone) hard rock
Shale
Plunge pool
River Tees
Erosion of softer shale below hard rock causes undercutting

Key words

- rapids
- waterfall
- estuary
- floodplain
- meander
- oxbow lake

Deltas

Some rivers have a **delta**. This forms when silt accumulates at the mouth of the river, especially during flooding, and the sea is unable to remove it. There are three types of delta:

- **arcuate** or **fan-shaped**, e.g. the Niger (Nigeria)
- **cuspate**, e.g. the Ebro (Spain)
- **bird's foot**, e.g. the Mississippi (USA)

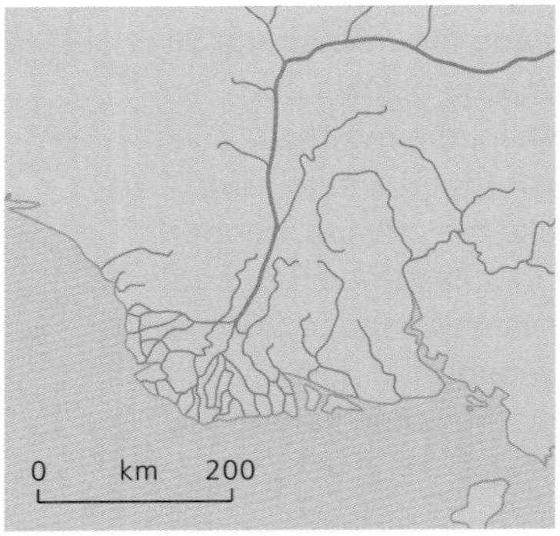

Niger: arcuate delta

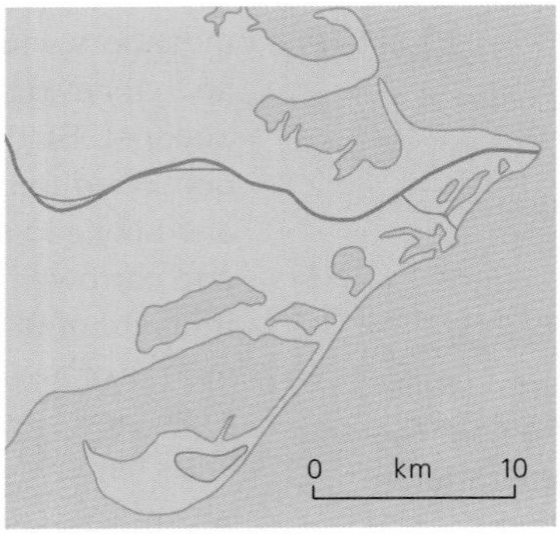

Ebro: cuspate delta

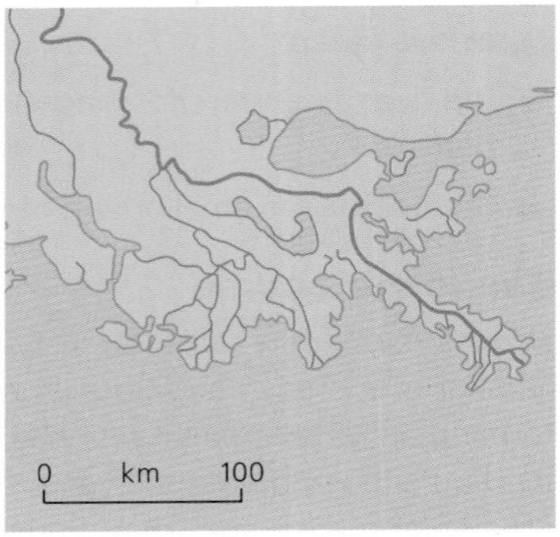

0 km 100

Mississippi: bird's foot delta

How the Nile Delta has changed

Since the completion of the Aswan High Dam in 1970 huge quantities of silt have been trapped behind the dam in Lake Nasser. Consequently, not enough silt reaches the delta. This has the following serious consequences:

- The delta is getting smaller and people have lost their homes and land.
- Valuable nutrients are lost and fish stocks have gone down.
- The soil is less fertile without the annual replenishment of silt. Farmers must now buy artificial fertilisers for their land. These are expensive and can cause pollution.

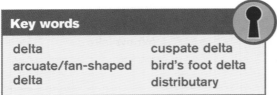

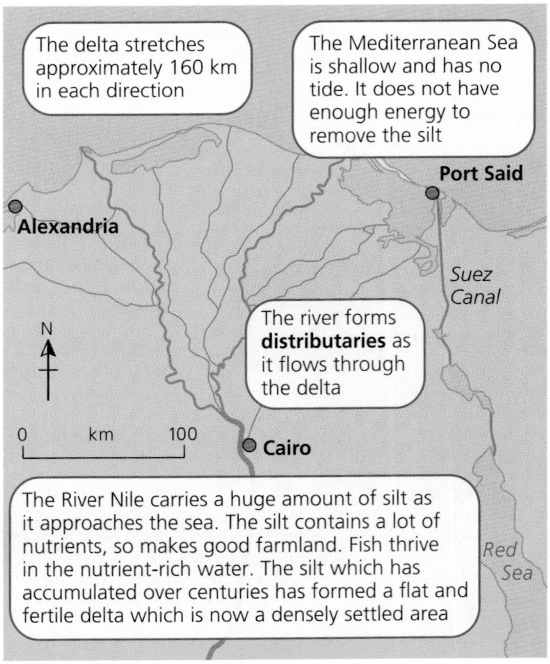

The delta stretches approximately 160 km in each direction
The Mediterranean Sea is shallow and has no tide. It does not have enough energy to remove the silt
Port Said
Alexandria
Suez Canal
N
The river forms **distributaries** as it flows through the delta
0 km 100
Cairo
The River Nile carries a huge amount of silt as it approaches the sea. The silt contains a lot of nutrients, so makes good farmland. Fish thrive in the nutrient-rich water. The silt which has accumulated over centuries has formed a flat and fertile delta which is now a densely settled area
Red Sea

Key words

delta	cuspate delta
arcuate/fan-shaped delta	bird's foot delta
	distributary

Test yourself

1 **Which is the odd one out in each of these groups of words? Explain why in each case.**
precipitation, infiltration, clouds
groundwater flow, throughflow, surface runoff
baseflow, lag time, peak discharge
embankment, reservoir, canal
meander, delta, floodplain

2 **Draw a diagram to show how a waterfall is formed and annotate it fully.**

3 **True or false? Explain your answer in each case.**
Flooding is more likely in built-up areas.
Afforestation is an example of hard engineering.
The point where a tributary joins the main river is called a confluence.
The level of the water table rarely changes.
Saltation is another term for the suspended load in a river.
The River Nile and the River Mississippi both have large deltas.

Exam tip

If you are asked to write about a landform, you must choose a feature made on the land, e.g. a delta or a floodplain. Meanders and oxbow lakes are water features.

Examination question

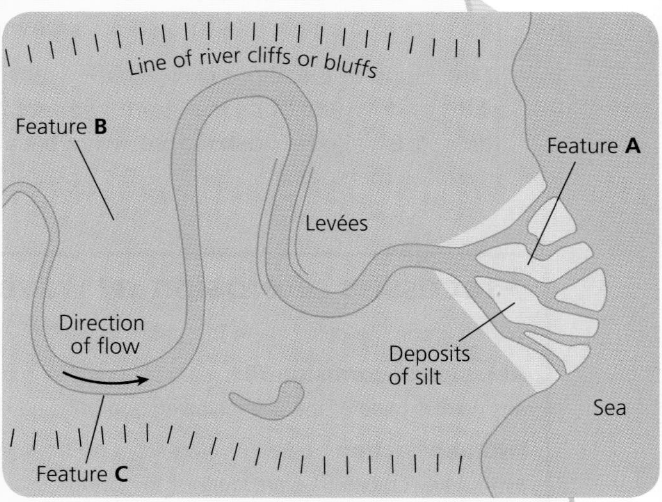

(a) **Name features A, B and C on the diagram.** *(3 marks)*

(b) **Explain the formation of an oxbow lake using diagrams to help you.** *(4 marks)*

Case study: *the causes, effects and management of river flooding*

Foundation tier:

(c) **(i)** **Name a river you have studied where flooding has caused problems.**

(ii) **Explain what caused the floods to occur.**

(iii) **Describe what steps have been taken to manage the flooding.** *(8 marks)*

Higher tier:

(c) **Name a river you have studied where flooding has caused problems.**
Describe how the river has been managed to reduce the flood risk.
To what extent are these protection measures sustainable? *(8 marks)*

Waves and erosion

Two basic factors affect the nature of the coastline: the waves and the type of rock.

Waves

The movement of water particles in a wave is circular.

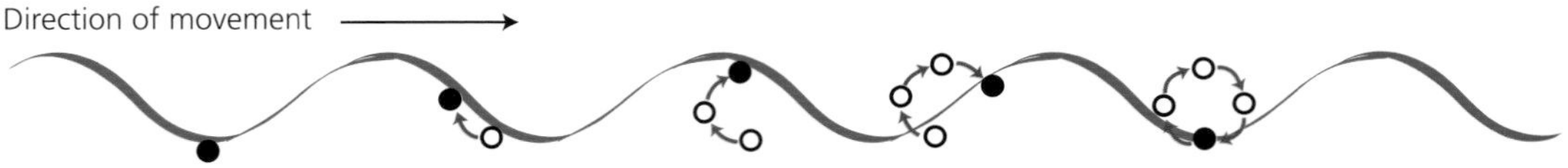

When the wave reaches the shore the circle is broken and the wave spills forward — it breaks.

If the slope of the shore is shallow, the wave spills forward for a long distance and is called a **constructive wave** because it pushes material onto the beach.

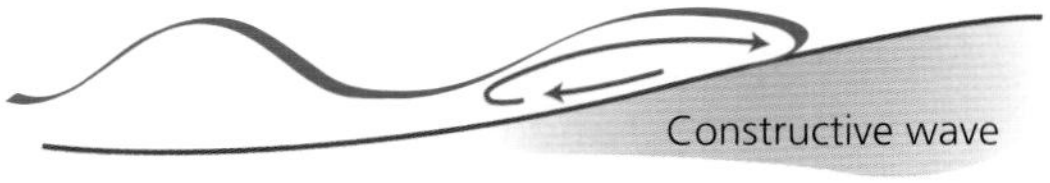

As the wave breaks, it swills up the beach. This is known as **swash**. It then runs straight back down the beach — known as **backwash**.

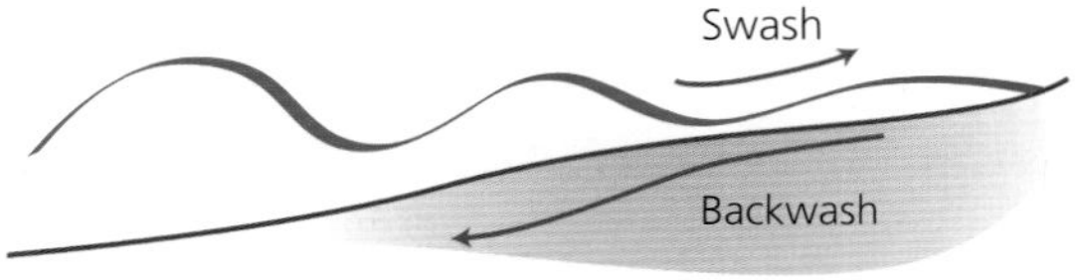

If the slope of the shore is steep, the wave plunges down and hits the shore with great force. It is called a **destructive wave** because it erodes the coast.

Processes of erosion by waves

Waves erode the coastline in four main ways.

Abrasion or **corrasion** The sea hurls pebbles and sand against the base of the cliff, chipping and grinding it down.

Hydraulic action Powerful waves lash the cliffs, forcing air into tiny cracks. The pressure of the compressed air weakens the rock and forces it to break up.

Corrosion The sea water reacts with chemicals and minerals in some rocks and they can be dissolved.

Attrition The rocks and stones which the sea erodes from the cliffs are rounded and broken down as they bump against each other and they are thrown against the cliff.

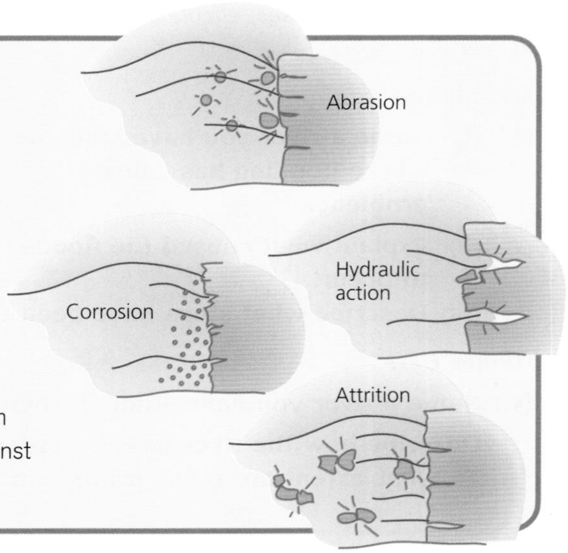

Weathering

Waves attack the base of the cliffs but the cliff face is likely to be eroded by weathering rather than by the sea itself.

In resistant rocks such as limestone and chalk water can get into cracks and fissures in the rock and widen them by chemical action or through freeze–thaw. Less resistant rocks can become waterlogged, causing landslips.

Freeze–thaw

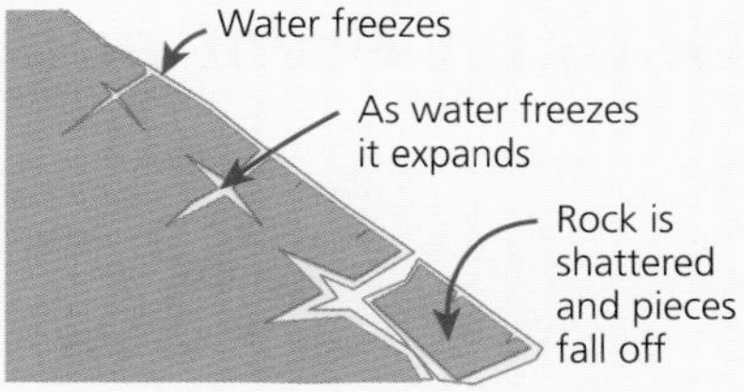
Water freezes
As water freezes it expands
Rock is shattered and pieces fall off

Erosion of cliffs

Hard rock cliffs such as chalk or limestone Cliffs formed from hard and resistant rocks such as chalk or limestone are eroded slowly. The cliffs are often high and almost vertical and erosion by the sea at the base of the cliff can cause rock falls.

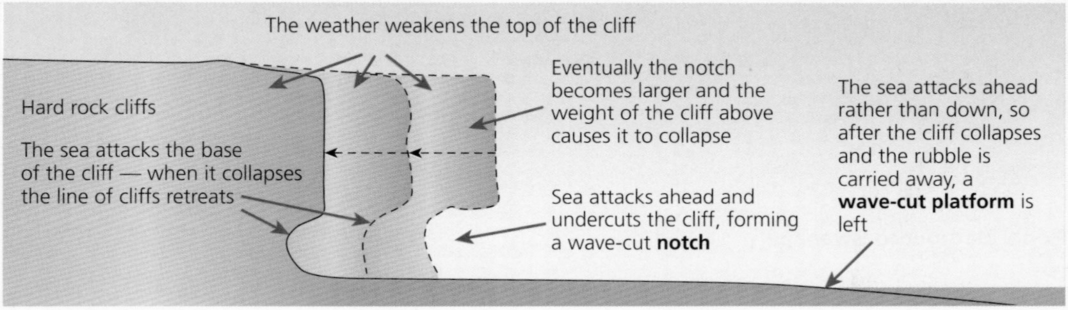
The weather weakens the top of the cliff
Hard rock cliffs
The sea attacks the base of the cliff — when it collapses the line of cliffs retreats
Eventually the notch becomes larger and the weight of the cliff above causes it to collapse
Sea attacks ahead and undercuts the cliff, forming a wave-cut **notch**
The sea attacks ahead rather than down, so after the cliff collapses and the rubble is carried away, a **wave-cut platform** is left

Soft rock cliffs such as sands and clay

(a)

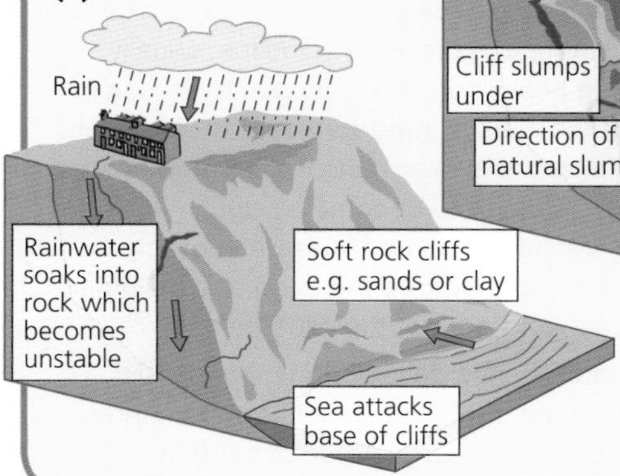
Rain
Rainwater soaks into rock which becomes unstable
Soft rock cliffs e.g. sands or clay
Sea attacks base of cliffs

(b)

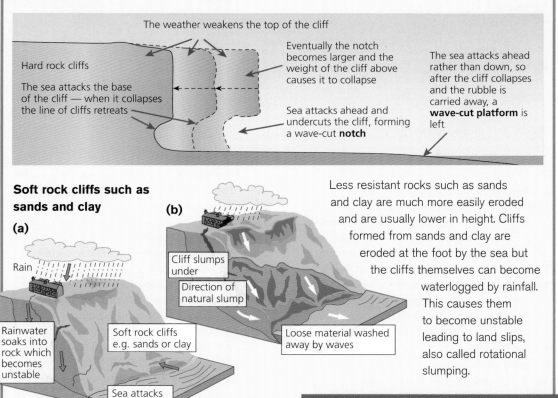

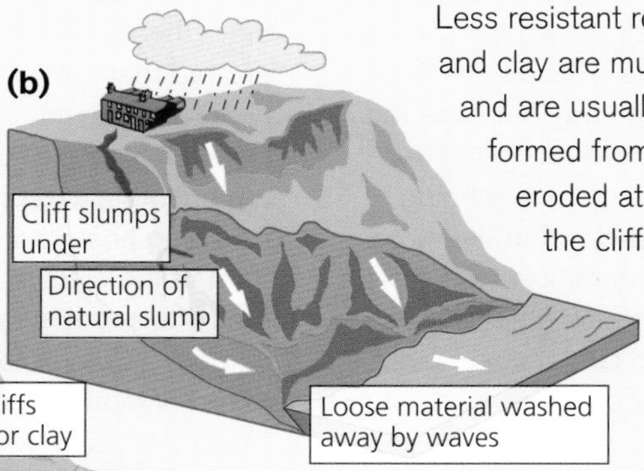
Cliff slumps under
Direction of natural slump
Loose material washed away by waves

Less resistant rocks such as sands and clay are much more easily eroded and are usually lower in height. Cliffs formed from sands and clay are eroded at the foot by the sea but the cliffs themselves can become waterlogged by rainfall. This causes them to become unstable leading to land slips, also called rotational slumping.

Exam tip

Practise drawing and labelling these diagrams so you can reproduce them in the exam.

Key words

constructive wave	hydraulic action
swash	corrosion
backwash	attrition
destructive wave	notch
abrasion/corrasion	wave-cut platform

Case study: *a coastal area and its landforms*

Handfast Point and Old Harry

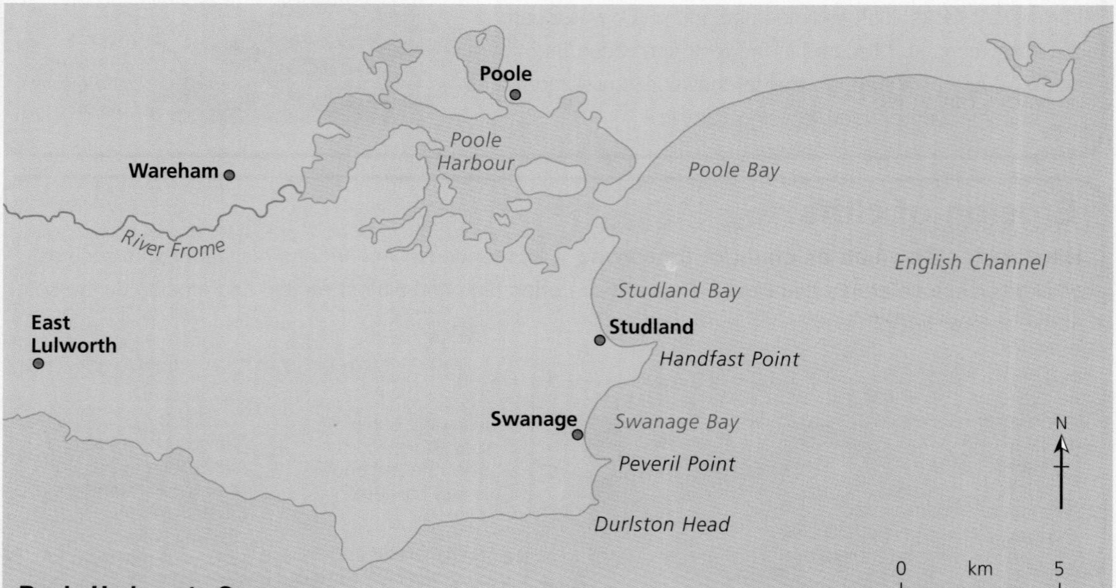

Poole Harbour to Swanage

The stretch of coastline form Poole Harbour to Swanage in Dorset is well known for its distinctive coastal landforms. This coastline is strongly influenced by its geology because the rocks lie at right angles to the sea. The softer clay and sands have eroded quickly, forming Swanage Bay and Studland Bay, but the more resistant chalk and limestone rocks are eroded much more slowly, forming headlands such as Handfast Point and Durlstone Head. A headland is a steep rocky promontory jutting out into the sea.

At Handfast Point the resistant chalk forms high cliffs, which are eroded slowly. The foot of the cliff is attacked by the sea (by hydraulic action, abrasion and attrition) and the face of the cliff is weathered by sun, rain and frost. Over many years erosion of the headland has led to the development of caves, arches, stacks and stumps. The largest stack at the end is called Old Harry.

Swanage Bay has developed where clay rocks meet the sea. These rocks are more easily eroded than the chalk and limestone headlands on either side. The clay has formed low cliffs at the back of Swanage bay which are unstable and prone to landslips, usually following heavy rain. The wide sandy beach in front of the cliffs is important as protection for the cliffs and it also attracts visitors and has helped Swanage to become a popular seaside resort. Wooden groynes have been erected along the beach to trap the sand and maintain the beach and protect the cliffs behind.

Studland Bay has developed where sands and clays meet the sea. This is a wide sandy beach backed by large areas of sand dunes. It is very popular with visitors.

Handfast Point and Old Harry

Old Harry is the name given to the large stack at the end of Handfast Point, the smaller stack or stump beside it is called Old Harry's Wife. The island between Old Harry and the mainland is called No Man's Land. These rocks were once part of the headland but erosion has removed the surrounding chalk. Although chalk is a hard rock it contains numerous cracks and fissures which are susceptible to erosion by the sea and the weather.

At the base of Old Harry and the other stacks is a wave-cut platform which is all that is left of the headland that once stood here. This is what remains after the cliffs collapse and the rubble is carried away.

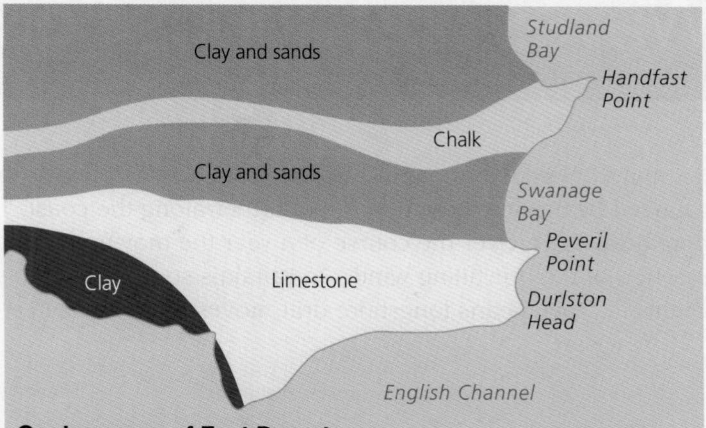

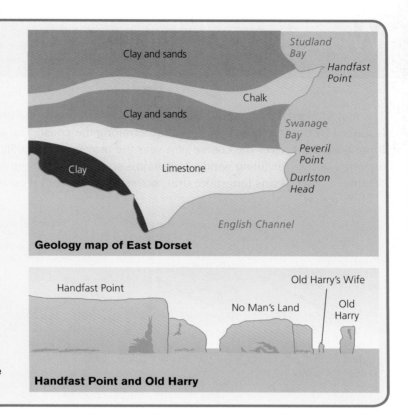

Geology map of East Dorset

Handfast Point and Old Harry

Erosion of a headland

Where a headland develops the sea can attack from three sides.

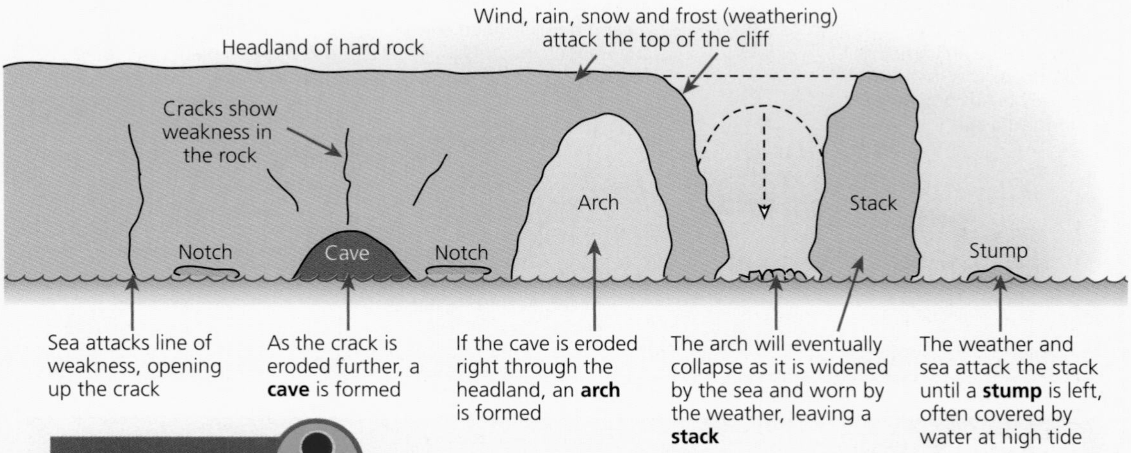

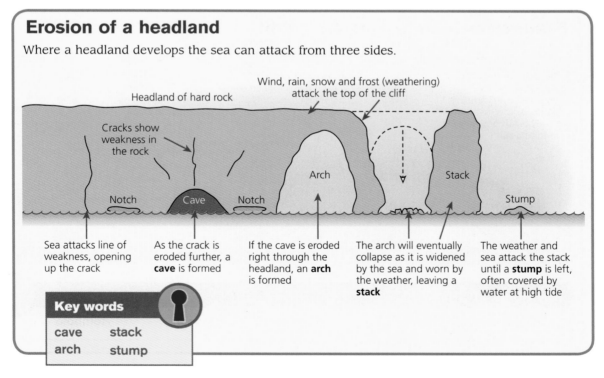

Key words

cave, stack, arch, stump

Waves and deposition

The material eroded from cliffs by the sea is worn down by the process of attrition and is moved by the sea to be deposited further along the coast. The sea moves material by **longshore drift**. Over the course of a year the movement of sediment will depend on the direction of the prevailing wind. On Britain's south coast the prevailing wind is from the southwest; this means longshore drift moves material from west to east.

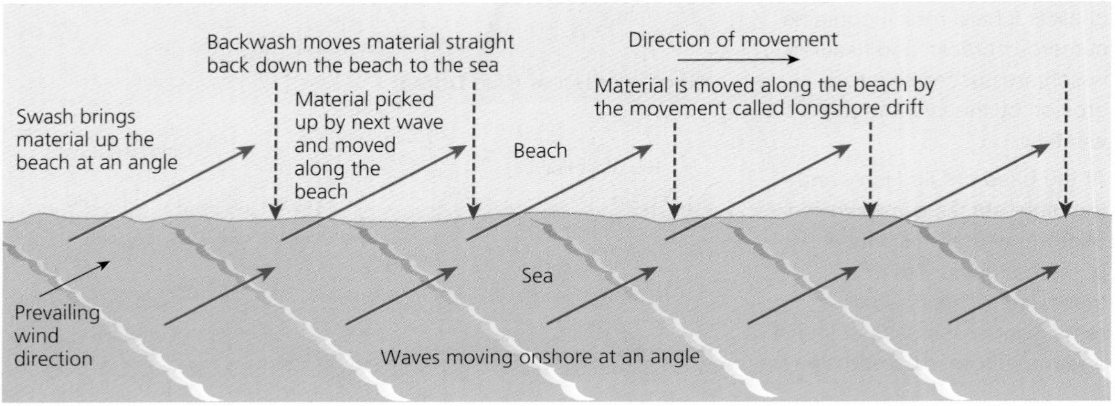

Features created by longshore drift

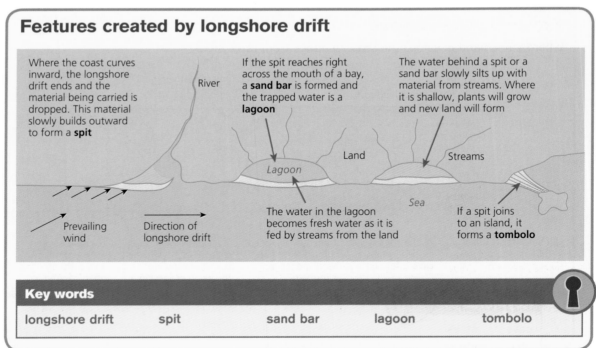

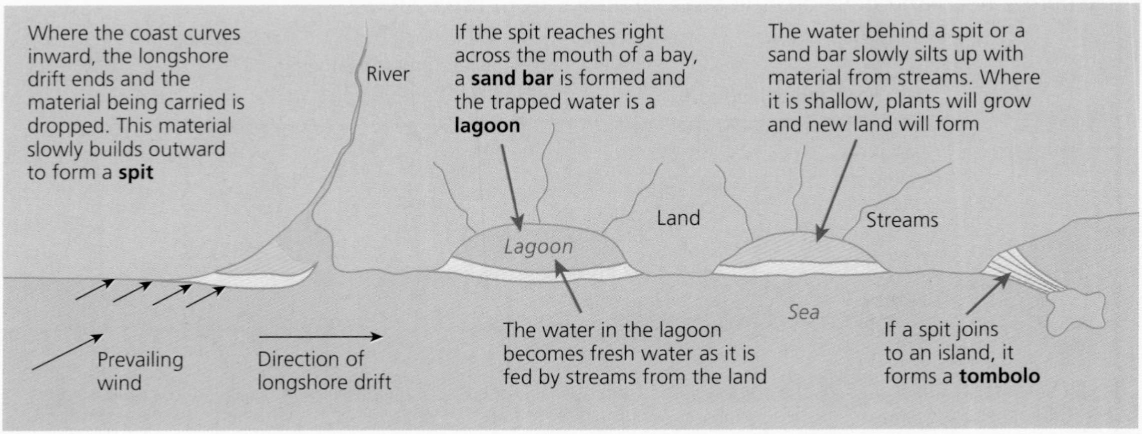

Where the coast curves inward, the longshore drift ends and the material being carried is dropped. This material slowly builds outward to form a **spit**
River
If the spit reaches right across the mouth of a bay, a **sand bar** is formed and the trapped water is a **lagoon**
The water behind a spit or a sand bar slowly silts up with material from streams. Where it is shallow, plants will grow and new land will form
Land
Streams
Lagoon
Sea
Prevailing wind
Direction of longshore drift
The water in the lagoon becomes fresh water as it is fed by streams from the land
If a spit joins to an island, it forms a **tombolo**

Key words

longshore drift | spit | sand bar | lagoon | tombolo

In Britain the coastline is constantly changing because erosion, transport and deposition are occurring. Land disappears in some areas due to erosion. Waves transport sediment along the coast and this may result in new land being created through deposition.

Exam tip

You may be asked to label a diagram showing how sediment is transported along the coast by waves. Practise drawing and labelling this diagram so you can reproduce it in the exam. Marks will be awarded for the accuracy and completeness of your labelling.

Spurn Head

Spurn Head is a narrow sand spit on the Yorkshire coast which extends across the Humber Estuary. It is 4.8 kilometres long (3 miles) but less than 50 metres wide in places. It has been owned since 1960 by the Yorkshire Wildlife Trust and is a designated National Nature Reserve and a Site of Special Scientific Interest (SSSI). It is an important location for migrant birds and many different types of wildlife.

Spurn Head

Plants such as marram grass bind the sand together but the spit is not stable and there is evidence that it has moved westwards in the past

The Humber lifeboat crew and river pilots live at the end of the spit. In stormy weather the spit can be breached, cutting off the tip and damaging the road which provides access to the houses

The damaged road has now been replaced with a movable road because the work of the lifeboat crew and river pilots is so important

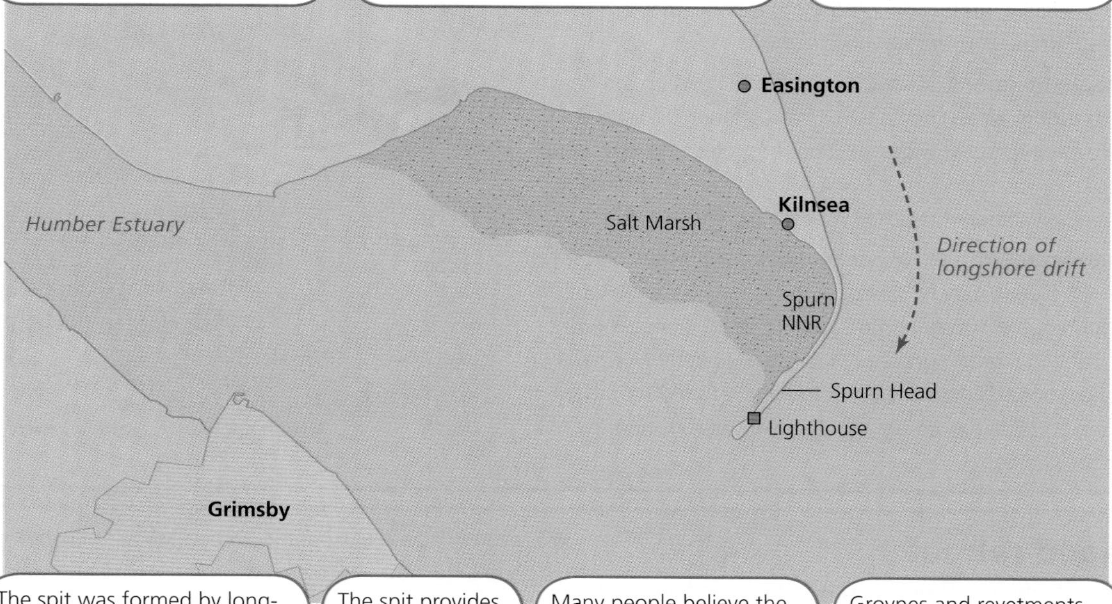

The spit was formed by long-shore drift moving sediment southwards along the Holder-ness coast and depositing it where the waters are sheltered by the Humber Estuary. The sediment has built up over centuries to form the spit

The spit provides shelter for the Humber Estuary

Many people believe the spit should be allowed to change position naturally as it has in the past; this is called managed retreat. This 'do nothing' approach is a cheaper and more sustainable option

Groynes and revetments were built on Spurn Head in the past to try to stop it being eroded. However they are no longer maintained because this is expensive and they are not able to prevent erosion

Coastal defences

Some sections of coastline need to be protected to prevent rapid erosion from the sea, however it is not possible to protect every stretch of vulnerable coastline because of the huge cost involved. Protecting one area of coast can result in other areas being eroded more quickly or being more likely to flood.

Where the sea is eroding the coastline it can do so at an alarming rate. This may not be seen as an issue when farmland is involved, but if homes and towns are threatened then it is more serious. The most common methods used to try to stop the erosion are hard engineering, soft engineering and **managed retreat**.

Key words

managed retreat	off-shore breakwater
groyne	beach replenishment
sea wall	dune management
revetment	integrated coastal zone management
gabion	
rip rap	

Soft engineering

These approaches are usually cheaper and do not damage the appearance of the coast. They are therefore a more sustainable approach to coastal protection. On the other hand, they are usually not as effective as hard engineering methods.

Beach replenishment Sand is brought in to build up the beach, either from further along the coast or from offshore. This looks completely natural and provides a beach to protect the coast and for visitors to enjoy. However, the sea will continue to erode the beach so replenishment has to be repeated every few years.

Dune management Sand dunes provide good natural protection for the coast. Dunes may be damaged by storms or by visitors walking through them to get onto a beach. Dune management includes planting marram grass to stabilise the sand or fill gaps and making wooden board walks for footpaths to reduce visitor impact. This protects the environment as well as reducing the risk of flooding and erosion.

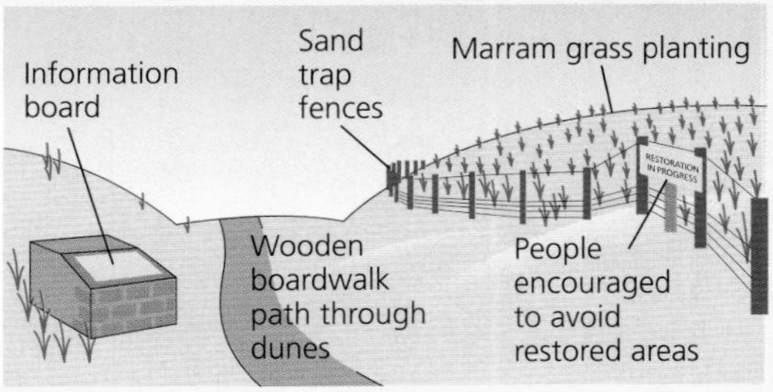

Managed retreat

In some places the sea is being allowed to erode the coast and people and activities have to move away. This is clearly the cheapest solution but it is very disruptive for the people who live where land and buildings are likely to be lost. In many cases compensation is not paid so individuals can lose a great deal of money. It can be very stressful and disruptive so has huge social costs. However this approach will result in an increase in the amount of salt marshes which help to stop future erosion and flooding and also provide a habitat for birds and wildlife.

Integrated coastal zone management (ICZM)

In many areas planners decide to combine hard engineering schemes with soft engineering and managed retreat. This kind of integrated approach is used on the Holderness coast in Yorkshire (see page 32).

Hard engineering

These structures are very expensive to build and maintain and are only used when towns, villages or expensive installations are at risk and where the economic benefit is greater than the costs involved.

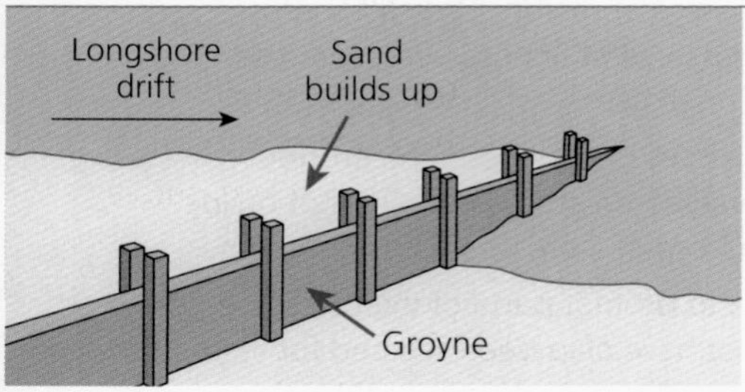

Groynes At seaside resorts wooden walls or groynes are built across the beach to stop the sand being washed away by longshore drift. The beach material builds up on one side of the groyne. Trapping the material like this may cause problems elsewhere as it stops the material moving down the coast where, for instance, it may be building up and protecting the base of a cliff. New groynes cost at least £200,000 each. They need to be maintained to stop the wood rotting.

Sea walls The most effective method of halting sea erosion. They are also the most expensive and cost about £500,000 per metre to build. Made of concrete, they are curved to deflect the power of the waves. But the sea can undermine them if the beach material in front of them is not maintained. Sea walls may by unsightly and also can restrict access to the beach.

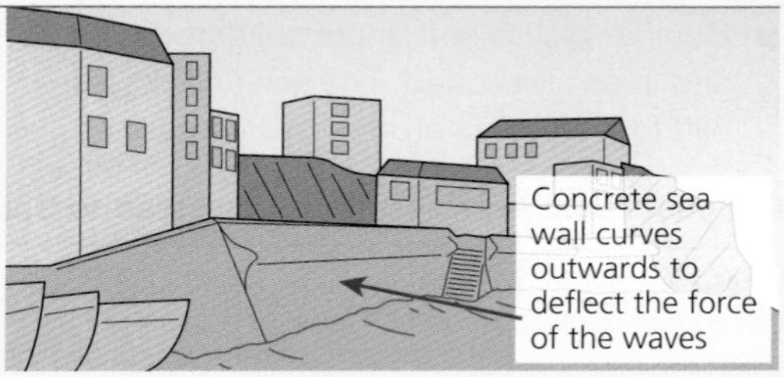

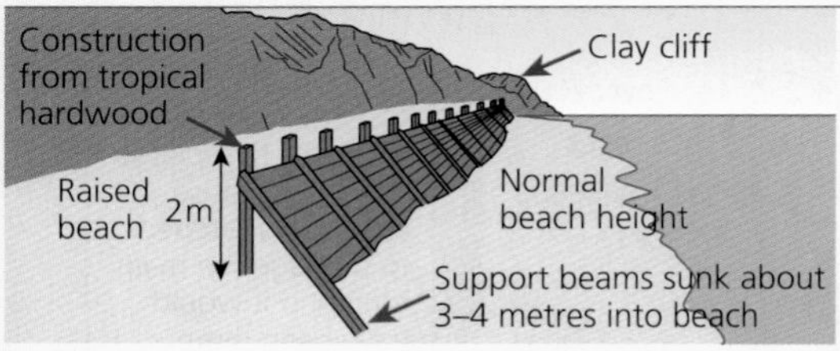

Revetments Sloping wooden fences with an open structure of planks to break the force of the waves and trap beach material behind them, protecting the base of the cliffs. They are cheaper but not as effective as sea walls.

Gabions Less expensive than a sea wall or a revetment. They are cages of boulders built up at the foot of the cliff or on a sea wall.

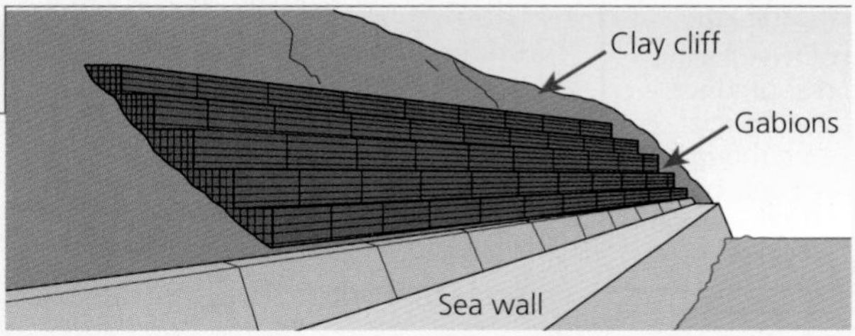

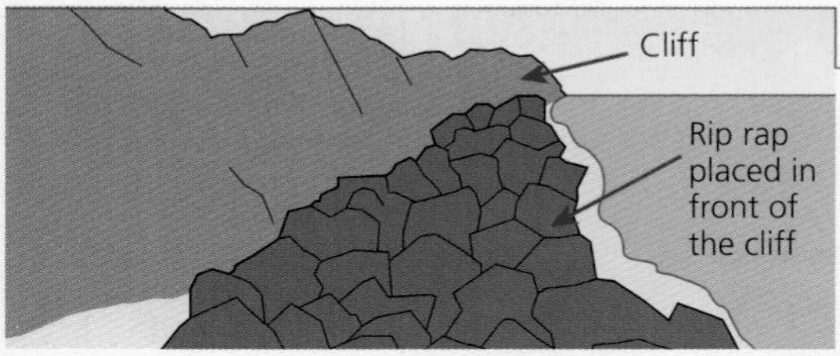

Rip rap The cheapest method but still expensive. Piles of large boulders are placed on the beach to protect the cliffs from the full force of the sea.

Off-shore breakwaters These are built on the sea bed a short distance from the coast and are usually made from rock or concrete. They are also very expensive to build but are effective because the waves break on the barrier before reaching the coast. This reduces wave energy and allows a beach to build up, which protects the cliffs.

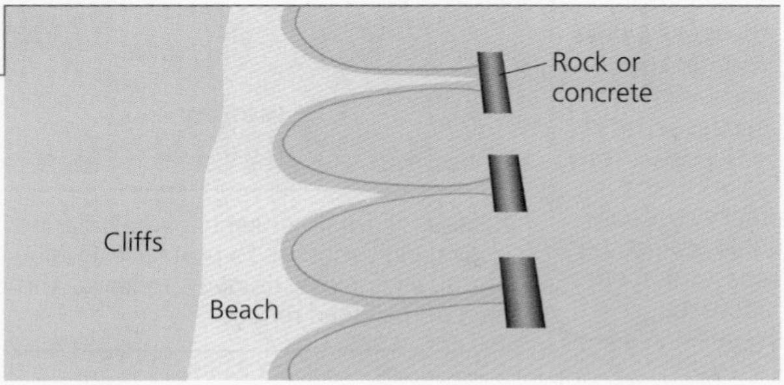

Case study: *coastline management*

The Holderness coast

The coast between Flamborough Head and Spurn Head is formed of boulder clay. This is a glacial deposit which is easily eroded. In the last 2,000 years 4 kilometres of land have been lost and this is Europe's fastest eroding coastline. On average 2 metres of land are lost each year.

There are two main reasons for this rapid erosion.

- Boulder clay is soft unconsolidated rock and is easily eroded. The sea attacks the cliff foot which is eroded by hydraulic action and abrasion. The cliff face is eroded by weathering, which causes rotational slip.
- Longshore drift sweeps the eroded material south so beaches do not build up in front of the cliffs to protect them.

Attempts to protect parts of the coastline in the past have also been blamed for increasing erosion in places further along the coast so the actions of people also contribute to the rapid erosion of this coastline.

The east coast from Flamborough Head to Spurn Head

Defences have been built at Easington to protect the main gas terminal for natural gas from the North Sea. It was decided that the benefit outweighs the cost of losing the terminal, even though it will only be safe for about 50 years

The three main settlements, Bridlington, Hornsea and Withernsea, are all protected by a combination of groynes, sea walls and revetments because the economic benefit of protecting these places outweighs the cost of the defences

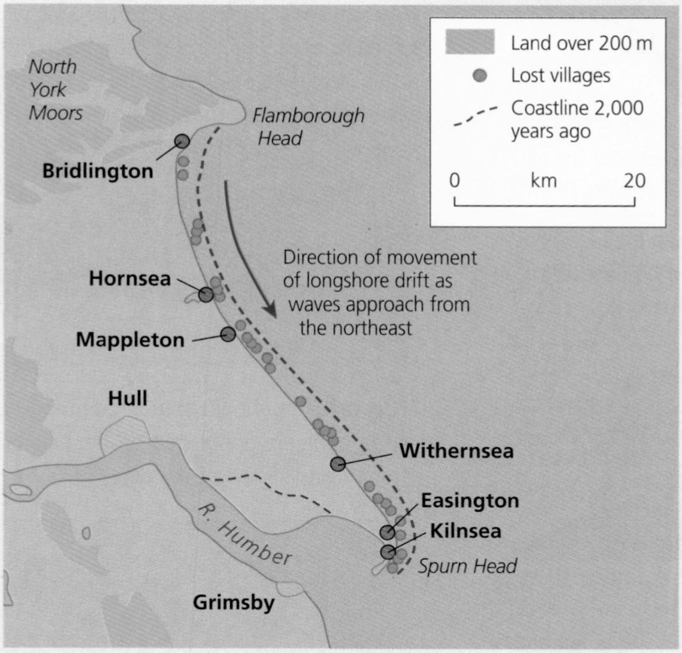

Sea defences were constructed at the village of Mappleton in 1991 because erosion threatened to damage the main road and it would have been more expensive to rebuild the road than the cost of protecting this section of coast. £2 million was spent on constructing two large rock groynes and a revetment. A beach has developed between the groynes which protects Mappleton and the road, however cliffs just south of the village have eroded even more quickly than before

These hard engineering schemes are expensive to build and to maintain, and as other parts of the coast retreat maintenance costs will increase. This is not a sustainable solution and in many cases will only protect the coast for about 50 years

Kilnsea is a small settlement which has no hard engineering schemes to protect it. Erosion will eventually force its residents to leave. This is an area of managed retreat

Many local people argue that planners should consider the social costs as well as the economic costs when making decisions about coastal management

Management of Holderness

In 2002 an integrated coastal zone management plan was produced for Holderness. This combines the use of hard engineering to protect some places with not protecting other places at all. This managed retreat means the unprotected areas will eventually disappear completely.

Everyone who lives on the Holderness coast would like to have sea defences but this is not possible as it would be far too expensive. Large settlements and places that are economically important are protected despite the high cost. Other places on the coast are not protected and the decision to 'do nothing' means the sea will continue to cause erosion. People who live in these areas of managed retreat, such as farmers, caravan site owners and householders, are losing their land and have to move further inland or move away completely when their land disappears. These people receive no government help and even have to pay for demolition costs of buildings themselves. Many people are very upset about this.

Mappleton

Mappleton is a small village with about 50 properties which lies about 3 kilometres south of Hornsea on the Holderness coast. The main road through the village (B1242) links all the settlements along the coast.

Mappleton was at risk of being lost to the sea because the coast was eroding so quickly.

The coast consists of boulder clay which was deposited by glaciers during the last ice age. It is a soft unconsolidated rock which is easily eroded both by the sea and by weathering.

On average 2 metres of land are lost each year but, in a stormy year, 7 to 10 metres of land can be lost.

A £2 million coastal defence scheme was approved in 1991. The risk of losing the vital road link justified this huge investment.

Two large rock groynes were built, together with a rock revetment along the base of the cliff. 60,000 tonnes of granite blocks came from Norway.

The groynes interrupt longshore drift, which carries sediment along the coast in a southerly direction. Sediment is trapped and has resulted in a wide sandy beach developing between the two groynes. The revetment prevents cliff-foot erosion, protecting Mappleton.

The rate of erosion south of the groynes has increased significantly because sediment being carried south is trapped by the groynes. This means there is no beach to protect the cliffs to the south which are now eroding at a faster rate than before.

Land and buildings have been lost at Great Cowden. The farmer here had to pay thousands of pounds for the demolition of his own farm buildings as the Council refused to pay compensation.

Mappleton

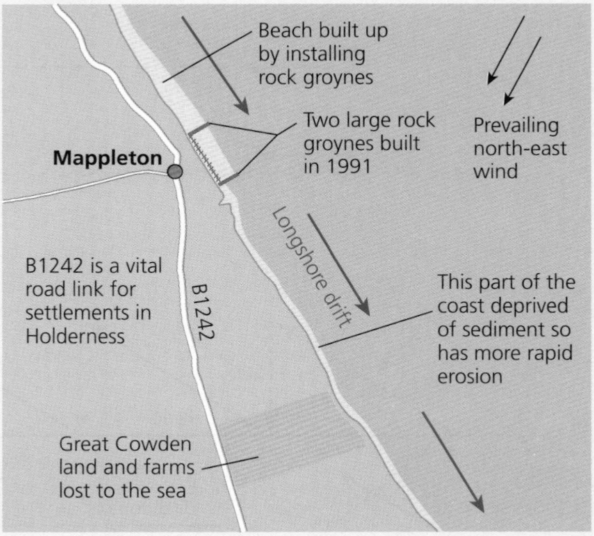
Beach built up by installing rock groynes
Two large rock groynes built in 1991
Prevailing north-east wind
Mappleton
Longshore drift
B1242 is a vital road link for settlements in Holderness
B1242
This part of the coast deprived of sediment so has more rapid erosion
Great Cowden land and farms lost to the sea

Test yourself

1 Tick those statements which are correct and put a cross next to those which are incorrect.

(a) A constructive wave spills forward on a gently-sloping coast, pushing material onto the beach. ☐

(b) Soft rock such as clay is easily eroded by the weather and the sea, so it forms headlands. ☐

(c) Corrosion is the process by which the sea hurls pebbles at the cliffs, wearing them down. ☐

(d) Attrition is the process by which rocks from the cliffs are rounded and broken down. ☐

(e) A wave-cut platform is left at the base of a cliff after the cliff has collapsed. ☐

(f) When a headland arch collapses a notch is left standing offshore. ☐

(g) The zigzag movement of the swash and backwash causes longshore drift. ☐

(h) If material builds across the mouth of a bay due to longshore drift it forms a sand spit. ☐

(i) Groynes are built to prevent longshore drift washing away the sand at a seaside resort. ☐

(j) Gabions are a more expensive method of preventing cliff erosion than revetments. ☐

2 Label the diagrams below correctly and give each a title.

Title: ....

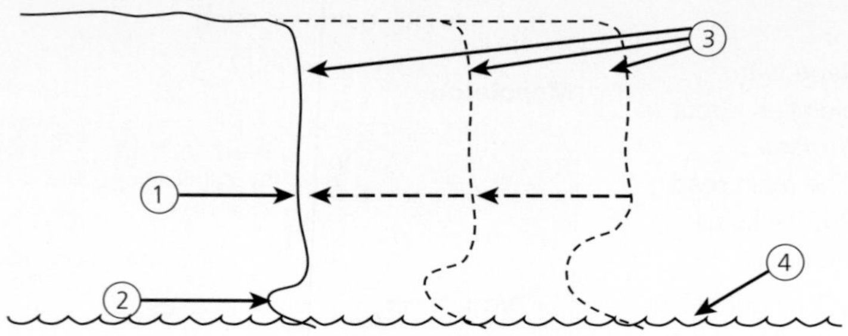

Title: ....

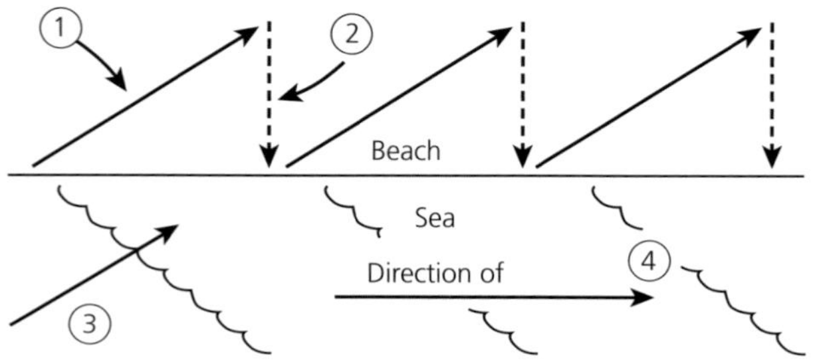

Examination question

(a) (i) **Label the diagram.** *(3 marks)*
(ii) **Describe two ways in which the sea wears away the coast.** *(2 marks)*
(iii) **Explain how the top of the cliff is worn away.** *(2 marks)*
(iv) **Explain why headlands are formed.** *(2 marks)*

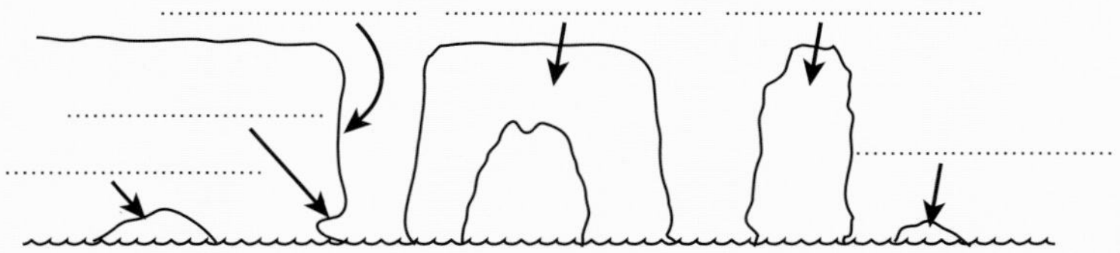

Exam tip

Make sure you have learnt and can use a case study for the final part of the question. The higher-tier question has three instructions or commands — can you find them?

Case study: *coastline management*

Foundation tier:

(b) (i) **Give a named example of a coastline where people have tried to stop the sea eroding the coast.**
(ii) **Describe how this coastline is being protected.**
(iii) **Explain the advantages and disadvantages of this scheme.** *(8 marks)*

Higher tier:

(b) Using a named example of a coastline which has been managed, explain how the coast has been protected. To what extent are these methods sustainable? *(8 marks)*

Notes

Theme 2

Population and settlement

Distribution of population

People are not evenly distributed over the Earth's surface. This is due to both physical and human factors.

Density of population is measured as the number of people per square kilometre. **Densely populated** areas have a large number of people per square kilometre.

Sparsely populated areas have a small number of people per square kilometre.

Note that most countries have both sparsely and densely populated areas. In the UK, examples are the sparsely populated Highlands of Scotland and the densely populated southeast England.

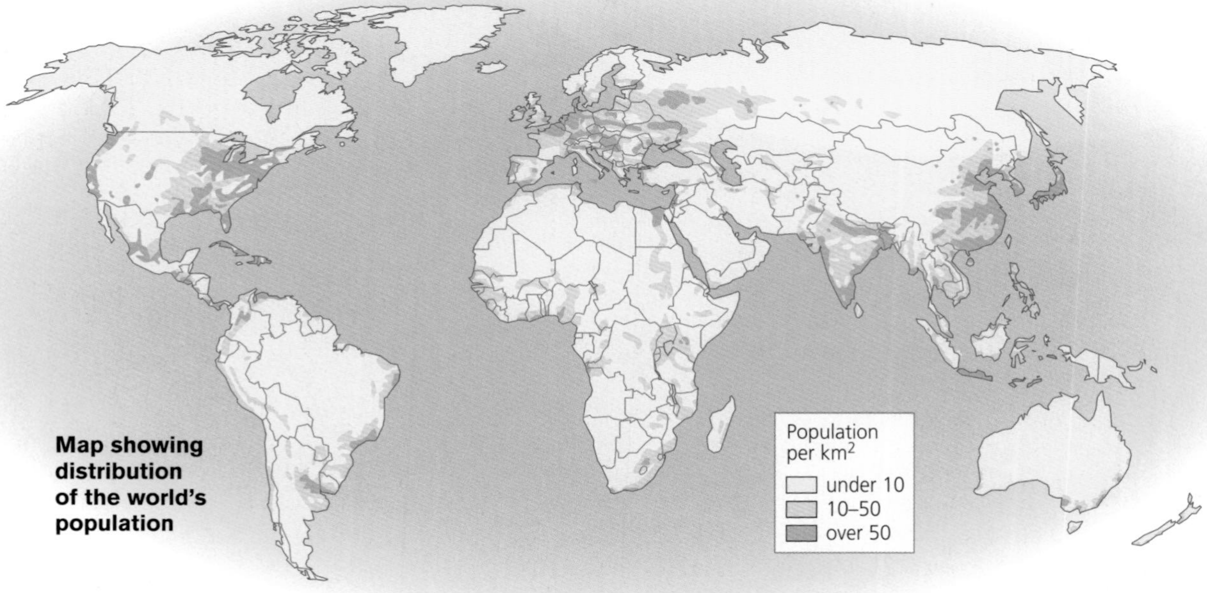

Map showing distribution of the world's population

Sparsely populated

70% of the Earth's surface is sea

Hostile environments:

- too cold, e.g. Arctic Canada
- too dry, e.g. Sahara Desert
- too mountainous, e.g. Himalayas

It is difficult to grow crops and keep animals, and life is uncomfortable

Densely populated

Areas with agreeable climates, e.g. California and Europe

Areas rich in natural resources such as coal, oil, gas and metal ores, e.g. Europe

Industrial areas where there is work, e.g. Japan, Western Europe, northeast USA

Fertile lowlands with sheltered valleys for growing crops, e.g. Ganges valley, southeast Asia

World population growth

- The world's population is growing rapidly.
- It took all of human history until 1804 to reach the first billion.
- The second billion took 123 years and the third billion took 32 years.
- In 12 years, from 1987 to 1999, the population grew from 5 billion to 6 billion.
- World population is expected to reach 7 billion in 2011. The latest prediction from **demographers** — the people who study population — is that the world population will level off at 8–10 billion, around 2050.

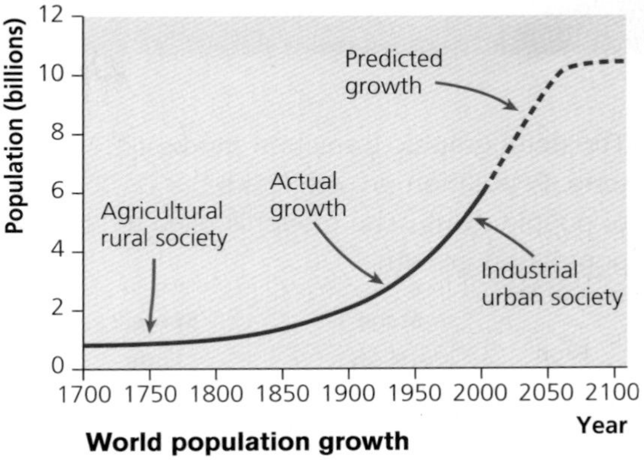

World population growth

Changing population growth

Population growth rates are different in different parts of the world. In general LEDCs, particularly countries in Asia and Africa, have increasing populations. Of all population growth since 1950, 95% has been in LEDCs and especially in the poorest LEDCs. This is because **birth rates** remain much higher than **death rates** so the rate of **natural increase** is high.

Birth rate is the number of live babies born per 1,000 of the population each year.

Death rate is the number of deaths per 1,000 of the population each year.

Natural increase is the difference between the birth rate and the death rate. If the birth rate is high and the death rate is low then the population will increase naturally. The natural increase is given as a percentage, although birth and death rates are expressed as a number out of 1,000.

Country	Birth rate (per 1,000)	Death rate (per 1,000)	Natural increase
India	39	13	2.7%
UK	13	9	0.4%

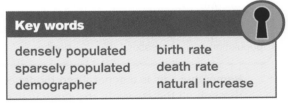

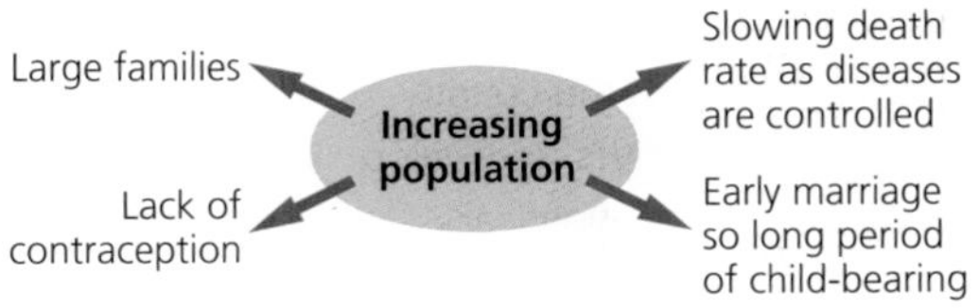

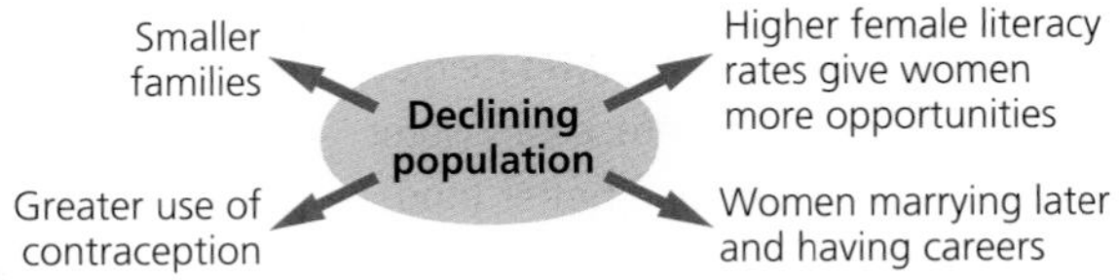

In most MEDCs populations are growing very slowly or even declining. This is because of improvements in housing, diet, education and medical care which help children to survive and people to live for longer. Falling death rates tend to encourage people to have fewer children so birth rates also fall. However some people continue to have large families for religious, economic or social reasons, while others who might want fewer children have no access to contraception or family planning advice.

In some countries, such as Italy, Portugal and countries of Eastern Europe, the population is decreasing and in many MEDCs, including the UK, the rate of growth is very slow.

Key words

densely populated	birth rate
sparsely populated	death rate
demographer	natural increase

The demographic transition model

The demographic transition model (DTM) shows how the growth of population changes over time. Population growth is a balance between the number of live babies born and the number of people dying. The model is used to show how countries pass through different phases of population growth.

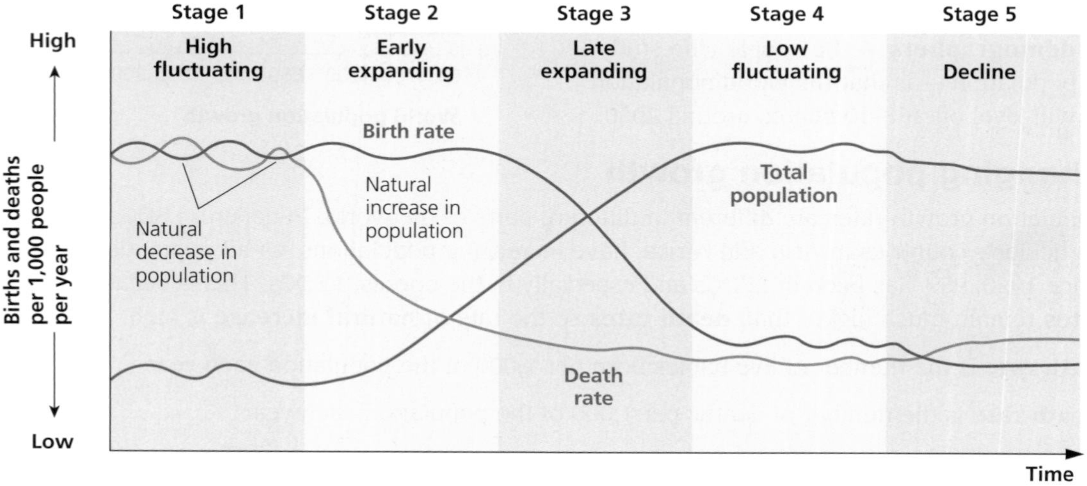

The demographic transition model

Stage 1 (high fluctuating) This shows a country or region before economic development. Both birth rates and death rates are high so there is a stable population.

Stage 2 (early expanding) As the country starts to develop economically death rates fall, because of improved medical care, clean water and better food supply. Birth rates remain high meaning that population increases.

Stage 3 (late expanding) Death rates continue to fall but birth rates begin to fall as well because continued economic growth leads to improved education, better access to contraception and family planning advice. There may also be changes in social attitudes. Population continues to increase but more slowly.

Stage 4 (low fluctuating) As living standards rise birth rates and death rates are both low, and in some cases there are fewer births than deaths. This reflects high levels of education and more women entering higher education. As more women work many choose to marry later and to have fewer children or even no children at all. This means population remains stable.

Stage 5 (decline) This stage has recently been reached by some countries. Death rates are slightly higher than birth rates. Modern medicine is keeping people alive longer, resulting in an ageing population. There are fewer people in the reproductive age range, so birth rates are low.

Remember this is a generalised model of population change. Not all countries will pass through all the stages in the same way or at the same speed.

Population structure

Population structure is the composition of a country's population by age and sex. It is usually shown as a **population pyramid** — which is like two bar graphs back to back, one for males and one for females. Age is shown in horizontal bars. The shape of population pyramids for different countries varies.

Life expectancy: average age people can expect to live.

Infant mortality: number of babies who die under the age of 5 years per 1,000 people.

Young dependants: children who are dependent on older economically active people.

Elderly dependants: people who are dependent on younger economically active people.

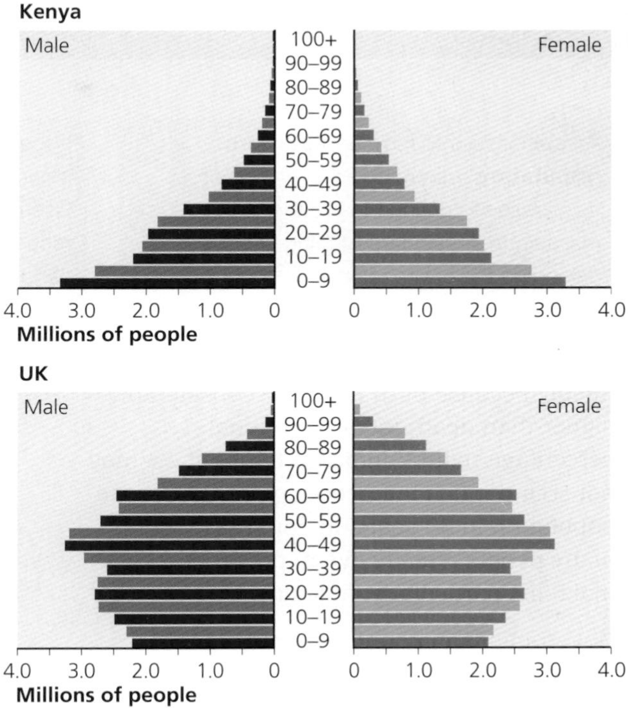

Population pyramids for Kenya and UK, 2009

The pyramid for Kenya shows many of the characteristics of LEDCs. The top of the pyramid is very narrow, showing there are few people over 65. There is a short life expectancy and few elderly dependants. The wide base shows a high birth rate and large numbers of children (young dependants). Because infant mortality and death rates are high the pyramid narrows in successive age groups. The high birth rate and rapidly growing population is a problem for many LEDCs such as Kenya.

The pyramid for the UK shows the characteristics of an MEDC. The narrower base shows low and falling birth rates so there are relatively few young dependants. The pyramid has a wide central part showing a large working population but also a wide top showing many people over 65, and more older women than men. This high number of elderly dependants is a growing problem for MEDCs such as the UK.

Population pyramids are used to help predict changes in the population and plan for the future. They can be used to predict the proportion of elderly people in the population who will need healthcare, or the number of young people who will be economically active in the future.

Key words

- population pyramid
- life expectancy
- infant mortality
- dependants

Population change can result in overpopulation

An area is said to have an **optimum population** if there are enough resources for people to continue to live there and maintain their standard of living without damaging the environment.

If there are too many people living in an area for the resources available then the area is **overpopulated**. Overpopulation is not sustainable so it is important to reduce population in order to increase the resources per person and improve standards of living.

In some LEDCs population is increasing rapidly because birth rates are considerably higher than death rates. An annual percentage natural increase of about 2% may not sound much but is enough to double the population in 30 years. In countries where many people already live in poverty this will make life particularly difficult.

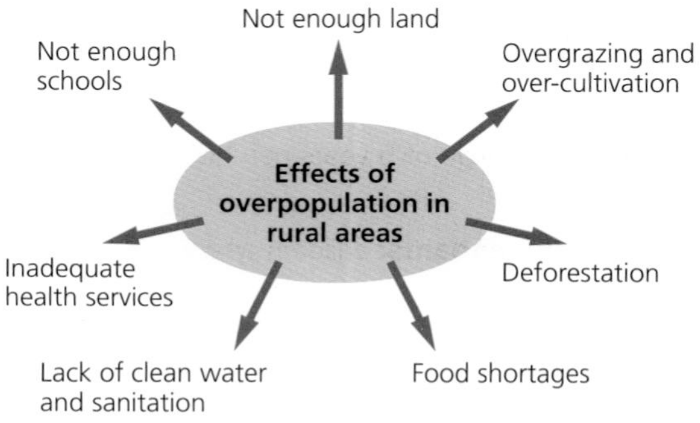

Governments in countries where the population is growing rapidly need to tackle the problem. This can be done in a number of ways, for example:

- restrict family size by law, as has been done through the one-child policy in China
- persuade people to have fewer children, for example by offering free contraceptives and family planning advice
- encourage people to marry later and to space out their children
- improve education, especially for girls, so they have a greater chance of finding paid employment
- improve healthcare so that infant mortality rates fall and children are more likely to survive to be adults

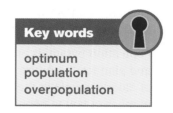

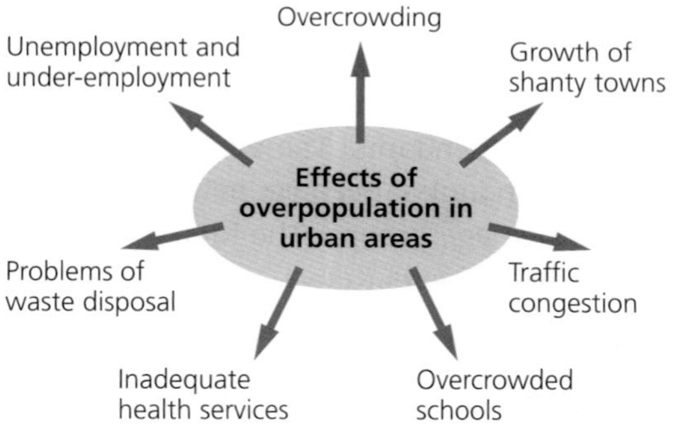

Many people believe the best way to manage a country's population is economic development which leads to improved education and healthcare and a higher standard of living. This is likely to be more sustainable than policies that force people to have fewer children.

Key words

optimum population
overpopulation

Case study: *an overpopulated country*

Bangladesh

Fact file

- Bangladesh is overpopulated; the population is too large for the resources of the country and there is widespread poverty.
- Bangladesh is smaller than the UK, but its population is two and a half times higher. It is one of the most densely populated countries in the world and has few natural resources other than fertile land; it has no fossil fuels or minerals.
- The population is over 160 million and has grown rapidly in the last 50 years from about 50 million in 1960. The population is still growing.
- Many women marry young and have had their first child before they are 19. Large families are normal and use of contraception is low.
- Over 30% of the population lives in extreme poverty on less than US$1 a day.
- Many families do not have enough to eat and about 30% of the population are undernourished, even though the country lies on flat, fertile land and has a warm subtropical climate allowing up to three crops a year.
- 80% of farms are less than one hectare in size because of the custom of dividing land equally between heirs. One hectare is too small to support a family.
- 25% of the population live in urban areas but almost three-quarters of these live in overcrowded shanty towns, many without access to sanitation.
- Less than half the population are literate and fewer women than men can read and write.

Strategies to manage population in Bangladesh

For many years governments have tried to limit population growth. The strategies aim to reduce the number of births by:

- providing family planning clinics (some of which are mobile clinics) which give advice about contraception and also provide health services for mothers and their children
- employing local literate women to go door to door giving out free contraceptives
- encouraging both men and women to be sterilised, with incentives such as food or money
- raising the legal age of marriage to 18 for women and 20 for men

Have the strategies been successful?

The average age of marriage for women has risen from 14 to 17 years and the proportion of women using contraception has also increased to about 50%. This has slowed the rate of population growth, but not enough, and Bangladesh is still predicted to have over 220 million people by 2050.

Critics say that for population rates to fall significantly and for this fall to be sustainable, much more needs to be done.

- Literacy and healthcare for women and girls must improve.
- Customs and traditions which expect girls to stay at home and look after the family need to change.
- More women should be educated and allowed to find paid employment because this means they are likely to marry later and have fewer children.
- Improvements to healthcare are needed so that fewer children die and therefore large families will no longer be necessary.

Population control

Many governments recognise the need to control their population in order to have a sustainable future. Mostly population policies aim to reduce birth rates by using persuasion and incentives, such as in India. However, in China there are strict laws about family size.

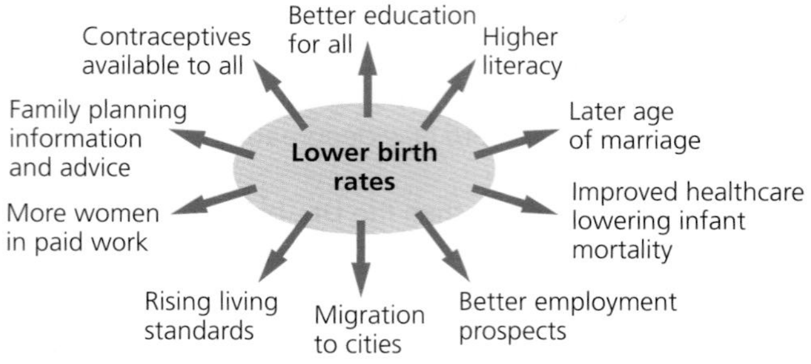

It is generally believed that economic development encourages people to have smaller families. This is because as a country becomes more prosperous there will be better healthcare and education as well as family planning advice, and access to contraception.

India

India has the second largest population in the world with over 1.1 billion people. Its population is growing quite rapidly and it is estimated that by 2050 India will have more people than China.

India has been trying to control its population for over 50 years. Despite some success it is still facing problems of overpopulation. For example, food production has tripled since 1950 but many people are undernourished. If the population is not controlled there will be food and water shortages and more people living in poverty.

In 2000 a national population policy was introduced to try to reduce population growth:

- contraception is free and available to all
- girls are encouraged to marry later (not before 20)
- healthcare is being improved to reduce infant mortality.
- it is compulsory for girls and boys to stay at school until they are 14

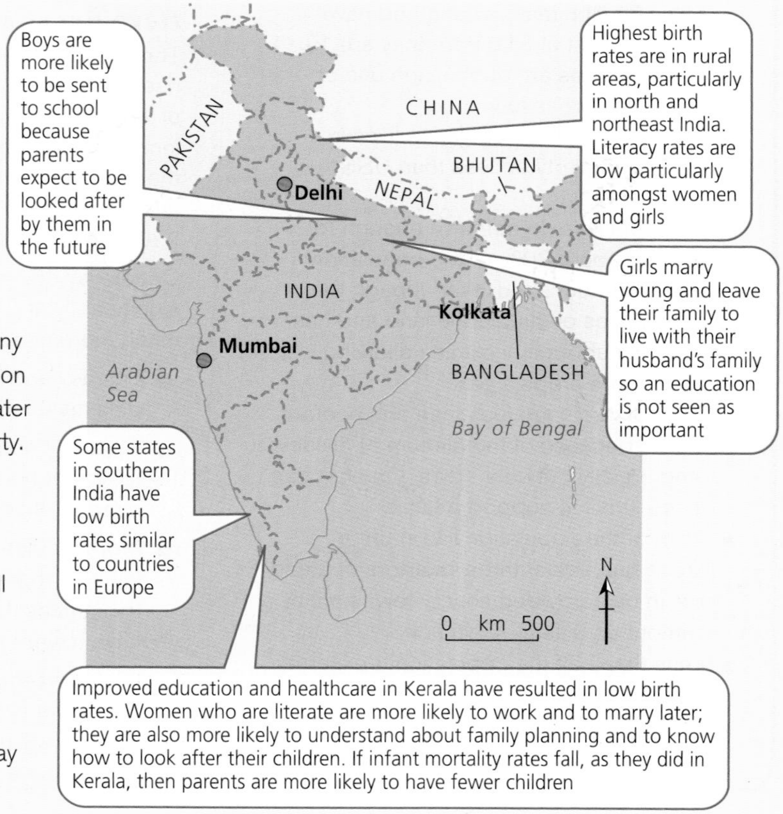

Boys are more likely to be sent to school because parents expect to be looked after by them in the future
Highest birth rates are in rural areas, particularly in north and northeast India. Literacy rates are low particularly amongst women and girls
Girls marry young and leave their family to live with their husband's family so an education is not seen as important
Some states in southern India have low birth rates similar to countries in Europe
PAKISTAN
CHINA
NEPAL
BHUTAN
Delhi
INDIA
Kolkata
Mumbai
BANGLADESH
Arabian Sea
Bay of Bengal
N
0 km 500
Improved education and healthcare in Kerala have resulted in low birth rates. Women who are literate are more likely to work and to marry later; they are also more likely to understand about family planning and to know how to look after their children. If infant mortality rates fall, as they did in Kerala, then parents are more likely to have fewer children

Case study: *a strategy to influence population change within a country*

China's one child policy

The People's Republic of China has the world's largest population at 1,331 billion in 2009, about 20% of the world's population.

China is the only country to impose rules about family size as a way of controlling population growth.

In 1979 the government introduced the 'one child' policy which stipulates that couples may only have one child. This was encouraged by giving preferential access to housing, schools and health services, and punished with fines on each additional child. Sanctions ranged from forced sterilization and pressure to abort pregnancy, to discrimination at work.

Today the policy is still in place but has been modified so that couples from an ethnic minority or where both are only children themselves may have two children. In rural areas, a couple may also have a second child.

Critics say:

The one child policy has led to forced abortions and forced sterilisation

There is evidence of female infanticide (killing girl babies) because of the traditional preference for boys. Men now outnumber women in China by more than 60 million

The same population reduction could have been achieved through voluntary means by encouraging couples to marry later and to space their children out, as was done between 1970 and 1979. Economic prosperity would also have resulted in lower birth rates

Many Chinese people are unhappy as they want a larger family

Many only children are the focus of attention in their family and have become very spoilt; this is called the 'little emperor effect'

The one child policy has been successful because:

- Population growth rates have fallen and there have been about 250 million fewer births than there would have been. It has also helped China's recent rapid economic growth.
- Problems resulting from overpopulation have been reduced, for example pressure on housing, healthcare, education and law enforcement.

The future

In 2008 a survey showed that over 76% of the Chinese population support the policy and the government said the policy would remain in place indefinitely.

The one child policy has slowed down population growth, however there are still over 6 million more births than deaths every year in China, which means the population is still increasing and is likely to do so until at least 2030.

As only children grow up they will have to support two parents and four grandparents; this is called the 4-2-1 problem. This is why if only children marry they are allowed to have two children of their own.

Young men may be unable to find wives as there are many more men than women.

Ageing populations

In many countries people are living longer and the proportion of older people is increasing, this is sometimes called the '**greying of the population**'. This is a problem because it means there are more older people who are dependent on a smaller **working population**.

In the EU, the proportion of people over 65 will nearly double by 2060, from 17% to 30%. At present there are four people of working age for each person over 65, by 2060 there will be only two.

In most EU countries the proportion of working people will decrease in the next 50 years, which will cause serious problems. For example, in Germany it is projected to fall by 30%. The working population in the UK is likely to rise, mainly because of the number of immigrants who lift the fertility rate.

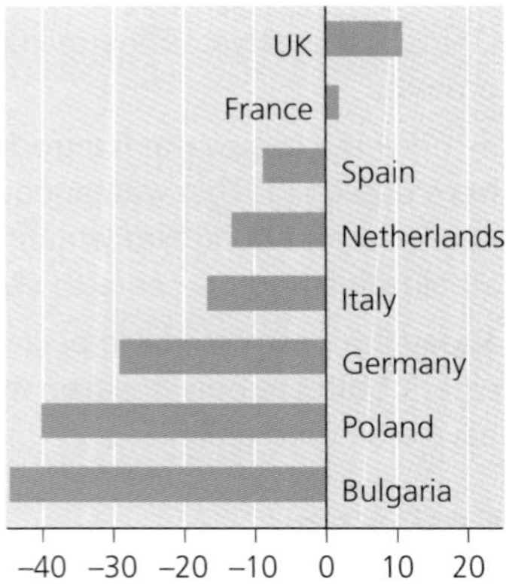

Estimated changes in the working population, 2008–2060 (%)

Reasons for ageing populations

- Increasing **life expectancy**: for example average life expectancy in the UK is 79 years.
- Low birth rates: in some countries birth rates are lower than death rates.

The consequences of ageing populations

- There is a growing market for leisure industries as older people enjoy spending money on holidays and days out.
- House prices in popular retirement places may rise.
- The cost of supporting older people through state pensions is increasing.
- There will be greater demand on medical services and long-term nursing care. In many countries healthcare is paid for by taxation from working people.

Solutions to ageing populations

Solutions are difficult to find but some ideas are:

- Raising the age of retirement so that older people work for longer and pensions are paid later.
- Raising taxes on the working population to pay for care of the elderly.
- Providing incentives to encourage people to have more children.
- Encouraging immigration of young skilled adults to fill the gaps in the labour market.

Raising taxes and not paying pensions until people are older are both unpopular but so too are policies which encourage immigration. Yet governments need to plan for the future as their populations get older.

Key words

life expectancy
greying population
working population
ageing population

Case study: *to illustrate strategies to influence natural population change*

Japan

Japan has the largest proportion of over 65s of any country (23%) and this is expected to rise to 30% by 2030. This is already causing difficulties as the number of working people declines and the problems are likely to get worse.

The population pyramids show changing population structure since 1950.

Falling birth rates give the pyramid an increasingly narrow base. The working population is also shrinking and the proportion of over 65s is increasing and is expected to be 35% by 2050.

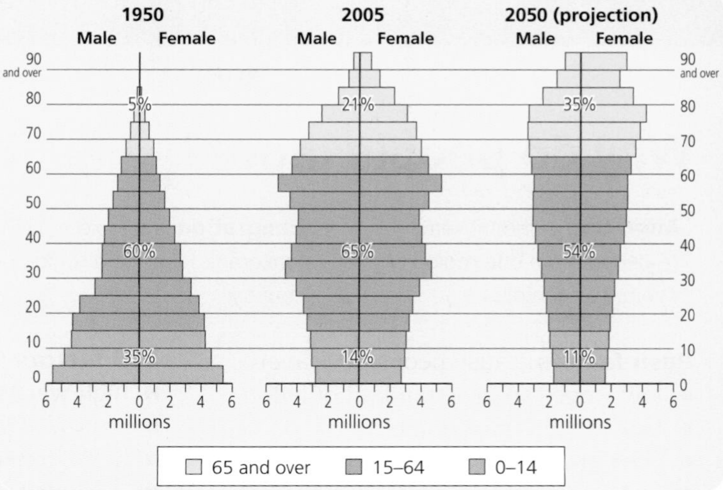

Japan's changing population

Reasons for the ageing population

- Japanese people have a life expectancy of 83 years. Good healthcare provision and a healthy lifestyle enable people to live longer.
- Average age of marriage is now between 28 and 30 years old. Marrying later means women are likely to have fewer children or even none at all. Some Japanese women are choosing to pursue a career instead of having children.
- It is the custom for children to be cared for by their mothers and few are sent to childcare facilities. Few women return to work after having children.
- It is very expensive to bring up children in Japan. Healthcare, school fees and university fees have to be paid by the parents so this puts people off having children.

Problems of the ageing population

- Japan is finding it difficult to fund people's pensions. The retirement age has already been raised from 55 to 60 and is now being raised again to 65.
- It will be increasingly difficult for the country to provide adequate healthcare to look after the large numbers of elderly people.
- In the future there will not be enough working age people to fill all the jobs. It is already difficult to find people to take dirty, dangerous or poorly paid employment.

What are the solutions?

- Unless population imbalance in Japan is tackled it will not be possible for the country to maintain its economic prosperity. Many people think the government has delayed making decisions about this for too long.
- The long-term aim is for Japanese women to have more children. The government is considering providing tax incentives to families and providing more childcare facilities, but this may not work without social and cultural change.
- The labour shortage could be solved by giving jobs to migrant workers but the Japanese as a nation are opposed to immigration and foreigners are not generally accepted. It will be difficult to change these strongly held views.

Migration

Why do people move?

Migration is the movement of people from one region or country to another.

Emigration is people (emigrants) leaving a region or country.

Immigration is people (immigrants) entering a region or country.

Push factors (cause people to leave)
- low wages so low standard of living
- lack of job opportunities
- poor quality of life
- lack of amenities, e.g. hospitals, schools
- conflict, e.g. civil war, oppression
- natural hazards, e.g. volcano, drought

Pull factors (attract people to an area)
- high wages and improved standard of living
- improved job opportunities, promotion
- better amenities and services
- improved quality of life
- better environment, no natural hazards
- freedom from oppression

There is usually more than one factor involved in the decision to migrate and there is usually a combination of both push and pull factors. Potential migrants also face a number of intervening obstacles that might prevent them moving, such as needing a visa to enter another country, not being able to speak the language or not having enough money to travel.

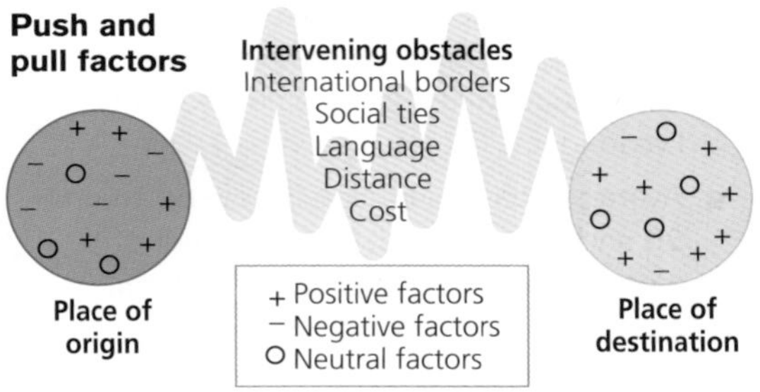

Case study: *international migration*

Migration of EU workers into the UK

The European Union (EU) encourages freedom of movement for workers between member states. This entitles UK citizens to move to other countries in the EU to work and allows EU workers to move to the UK.

In 2004 ten new countries joined the EU, including Poland and the Czech Republic. No restrictions were placed on people moving to the UK and within two years more than 400,000 migrant workers had arrived. In 2007 when Bulgaria and Romania joined the EU the government placed restrictions on the numbers of migrant workers entering the UK.

Who migrates?

- Mostly young and single people many of whom do not intend to stay permanently.
- Many skilled, well educated people who cannot find jobs in their own country.
- Some professional people including doctors and dentists.

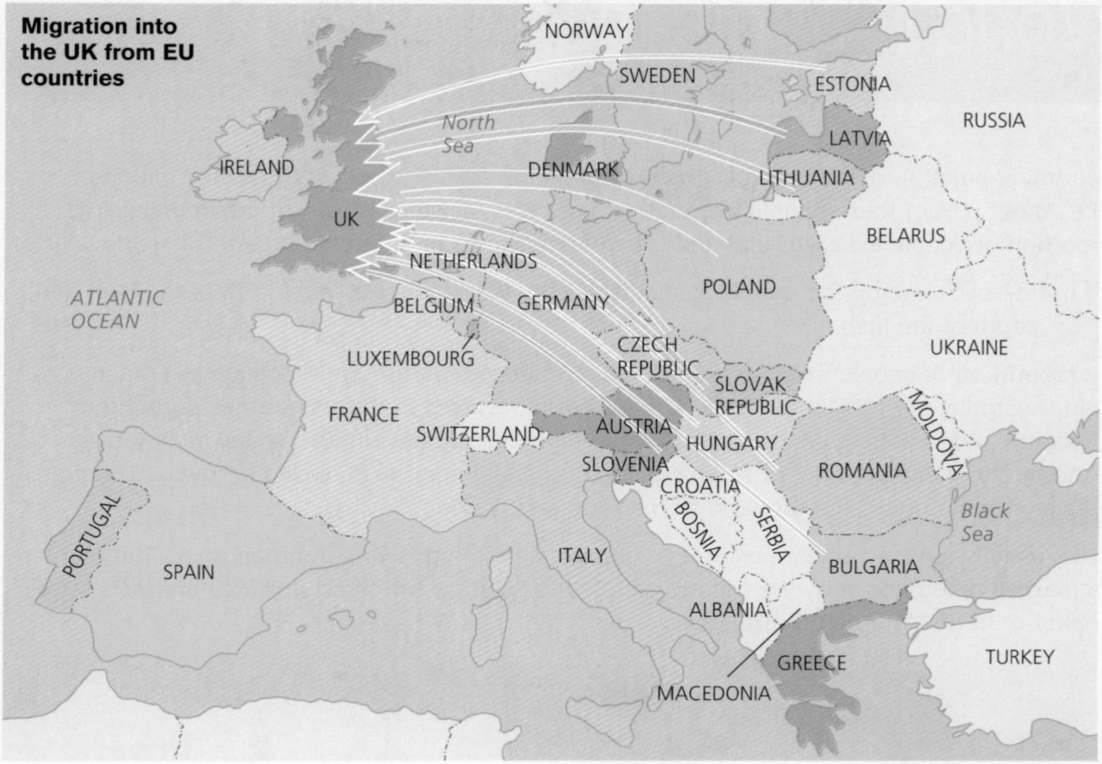

Benefits to the UK

- Helps employers to find workers when local people are not available or not willing to work, for example care workers, farm workers, builders and bus drivers.
- Local economies are boosted because migrants spend money in local shops and services. Councils can collect more council tax to invest in the local area.
- Most migrants pay tax which contributes to services such as the health service, education and transport. As the majority of migrants are young and single they usually make little use of these services themselves.

Problems

- There have been social problems in some areas where local people resent large numbers of newcomers. There may be pressure on rented properties and other services in the short term.
- Migrant workers often take low paid jobs which means unskilled and poorly qualified UK workers may lose out.

The future

- It is estimated that nearly 1 million workers have entered the UK and about half a million are still here. However, there are now fewer new arrivals and they are counterbalanced by those returning home.
- Recession and the weak pound compared with the Euro mean the UK is no longer such an attractive place for potential migrant workers.
- Some multinational companies (MNCs) have closed factories in the UK and concentrated production in factories in countries that have recently joined the EU, which will mean there are likely to be fewer migrants into the UK in the future.

Key words

- migration
- emigration
- immigration
- push factor
- pull factor

World urbanisation

The total population of the world is now over 6 billion and is expected to reach 7 billion in 2011. About 50% of these people live in towns or cities, in other words, in urban areas. The proportion of people living in rural and urban areas varies in different parts of the world.

The north–south line on the map roughly divides the world into MEDCs (north of the line) and LEDCs (south of the line).

The proportion of people living in urban areas is much greater in MEDCs, such as France, Japan, Australia, the USA and the UK. Urbanisation occurred in these countries during the nineteenth and twentieth centuries, following the Industrial Revolution. Today most jobs in MEDCs are in manufacturing or service industries, which tend to be based in towns. There are relatively small numbers of farmers, so the rural population is low.

LEDCs usually have a smaller percentage of their populations living in urban areas, but there is a marked difference between countries in South America and those in Africa and Asia. More people live in towns in South America.

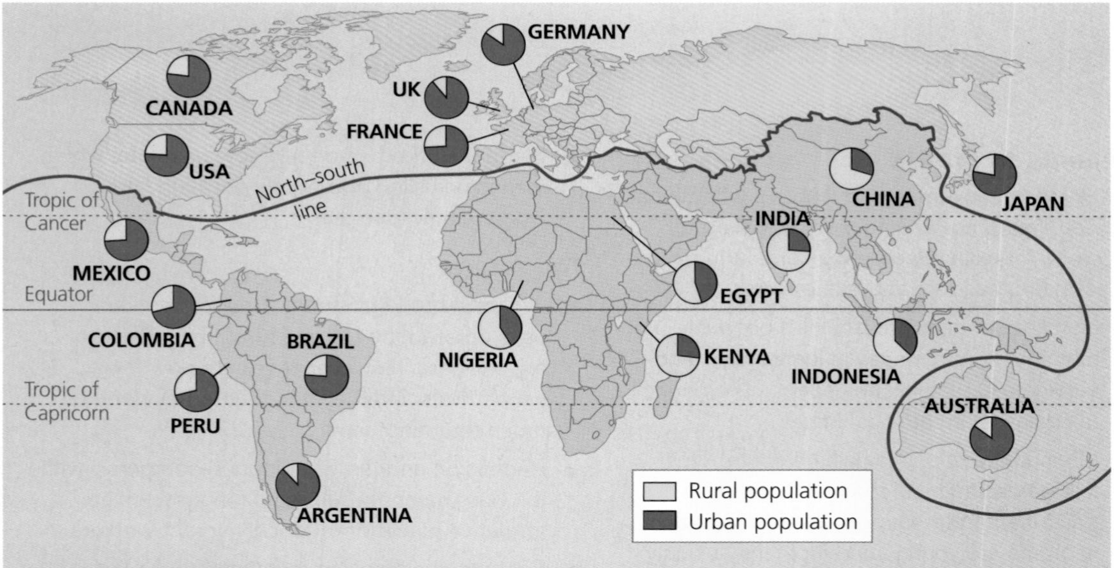

Urbanisation in selected countries

Urbanisation is the increase in the proportion of people who live in towns and cities. It is occurring on a global scale and in most LEDCs. It is no longer taking place in Europe and America.

Urban growth is the expansion of towns and cities so that they cover more land, as well as gaining larger populations.

Causes of urbanisation

Cities in LEDCs are growing rapidly. This is due to two important processes:

Natural population increase The growth in population because birth rates are higher than death rates. This is common in LEDC cities because many young people but few old people live there.

Rural-to-urban migration The movement of people away from the countryside into towns and cities. Rural to urban migration is the result of **push** and **pull factors**:

Pull factors attract people to the cities:

- the hope of work and the chance to make money
- better schools
- healthcare
- 'bright lights' and entertainment

Push factors encourage people to leave the villages:

- poverty
- few jobs except in farming
- poor health facilities
- few schools
- little entertainment, especially for young people

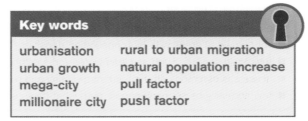

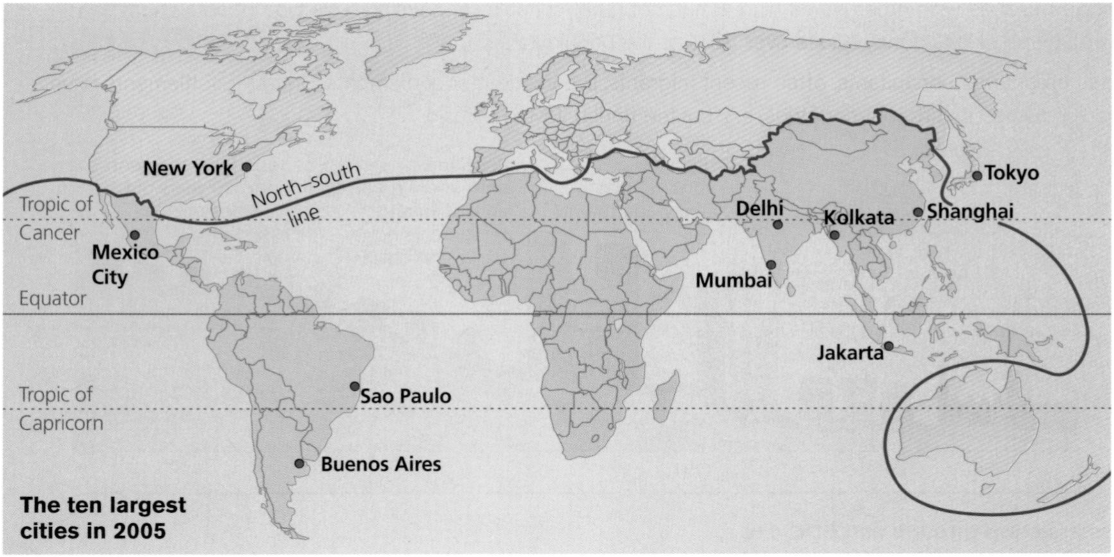

This map shows the ten largest cities in the world according to UN estimates. Look at their distribution. Only two of them, Tokyo and New York, are in MEDCs. All the rest are in LEDCs.

All these cities (and nine others) have over 10 million inhabitants and are known as **mega-cities**. Over 300 cities worldwide have more than 1 million inhabitants and are known as **millionaire cities**.

Key words

urbanisation	rural to urban migration
urban growth	natural population increase
mega-city	pull factor
millionaire city	push factor

Consequences of urbanisation

Urbanisation is helping LEDCs to develop economically:

- Industrial growth provides employment.
- New businesses and enterprises create more employment, benefiting the local economy.
- Cities also provide better services than most rural areas, particularly healthcare and education.

However, rapidly growing cities also have many problems, particularly in LEDCs where there may be little money available to invest in city planning.

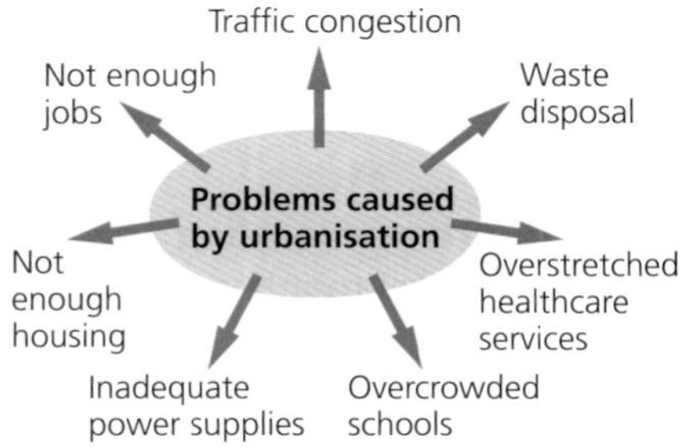

Shanty towns or squatter settlements

Rural–urban migration in LEDCs usually leads to the development of extensive **squatter settlements** or **shanty towns** on the outskirts of the city or on very steep slopes, land liable to flood or land near to quarries or large factories — in other words land that is not really suitable for housing. This type of settlement can cover a large area and may house over 30% of the population.

The poorest urban inhabitants, often recent migrants, live in squatter settlements. Squatter settlements have different names in different countries, e.g. bustee (India), favela (Brazil).

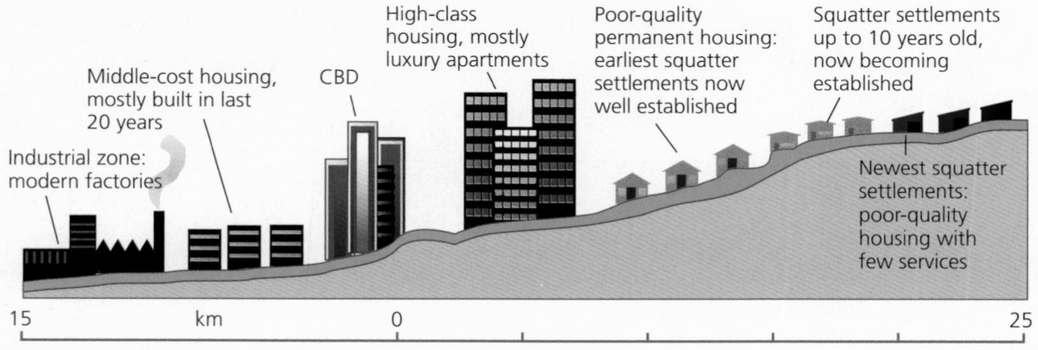

Cross-section through an LEDC city

Squatter settlements are built illegally by people who cannot afford proper housing — often newly arrived migrants. Houses are usually made with locally available materials, with mud bricks or wooden walls and corrugated-iron roofs.

The main problems are:

- very crowded areas with high population densities
- few basic services such as clean water and sanitation, waste disposal, electricity supplies or paved roads
- disease is common with high infant mortality and a short life expectancy
- few 'formal' jobs so many people earn what they can from 'informal' occupations such as street trading

Key words

squatter settlement
shanty town

Managing urbanisation

There are many problems to tackle in rapidly growing cities including traffic, waste disposal and pressure on services. In many cases it has been a priority to tackle the problems of the poorest housing.

New housing developments

City authorities are aware of the problems of large squatter settlements but rarely have enough resources to tackle them. In some cities, such as Lagos in Nigeria, the authorities have built apartment blocks to re-house people. Building new apartments is being planned for Mumbai in India but in many cities this is not an affordable option.

Self-help schemes

Once people have built a house, no matter how basic it is, they are likely to improve it when they can. However, they will only do this if they are confident they will not be thrown off the land. People must therefore be given legal ownership of the land.

Self-help schemes are important in most big cities in LEDCs. People improve their houses slowly, for example replacing mud walls with bricks or breeze blocks, and fitting proper windows and doors. The house may gradually be enlarged, building more rooms and then adding upper floors.

Later city authorities may provide clean water from standpipes in the street, and help with sanitation and waste collection. Bus operators will start bus services and health centres may be built. In this way people work together to improve their area and over time it changes from a poor, illegal settlement to a legal, medium-quality housing area.

Non-governmental organisations (NGOs) such as Oxfam and Save the Children encourage people to help themselves through **micro-credit schemes**. These schemes provide small grants or loans to the poorest people who not would otherwise be able to borrow money. NGOs provide training and advice and this, together with the loan, helps people to start their own business, for example running a market stall, hairdressers or car repair workshop.

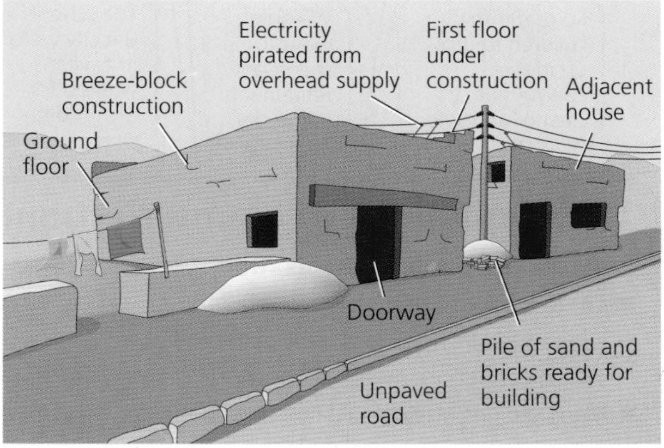
Electricity pirated from overhead supply
First floor under construction
Breeze-block construction
Adjacent house
Ground floor
Doorway
Pile of sand and bricks ready for building
Unpaved road

Typical self-help scheme housing

Key words

- self-help scheme
- non-governmental organisation (NGO)
- micro-credit schemes

Counter-urbanisation

During the last 30 years many cities in MEDCs have actually lost population. People and businesses have moved away from large urban centres to small towns or villages. This process is known as **counter-urbanisation**. People move out of large cities for a number of reasons (push factors) and are attracted to villages or small towns by pull factors.

Some villages and country towns have grown rapidly in the last 30 years through **urban–rural migration**. New houses have been built and older ones renovated. In some places the total population has more than doubled, bringing with it great changes to the village, some good and some not so good. These **suburbanised villages** are usually within easy reach of a large town or city because many people who choose to move are commuters who travel back to the city each day to work

Counter-urbanisation is not just about commuters. People may choose to move out of the city when they retire, moving to more scenic areas such as a national park or the coast. Also changes in technology now allow some people to work from home and therefore have the freedom to live where they want to, which may be away from the city in a rural area.

Counter-urbanisation: push and pull factors

There's a good road and a rail service from here so we can easily get to the city if we want to

I'd prefer my children to grow up in the countryside. It's cleaner and safer

We didn't want to spend our retirement in the city

It's expensive in the city. House prices are high, especially in the nicer areas

The schools in the city were not what I wanted for my children

The city is too busy and noisy. There are endless traffic jams and there is nowhere to park

We've bought a nice house with a garden and we can walk to the shops or out into the countryside

I moved my business because I found an office with a good view and easy access. So much of our work is done by e-mail, fax and telephone that it doesn't really matter where our office is

Consequences of counter-urbanisation

Benefits of counter-urbanisation

- There are more people in the village to support local shops, bus services and the village school.
- Many old or derelict buildings have been renovated.
- There are opportunities for local businesses such as builders, shops, the garage or the pub, and jobs are created locally.
- Younger people will settle in the village, creating a more balanced population structure.

Problems caused by counter-urbanisation

- House prices go up, making it difficult for local people to buy a house.
- Open space is built on and green belt land may be lost.
- There are more cars, causing traffic and parking problems and creating a risk for pedestrians.
- Main roads into the city can become congested as people commute to work daily.
- Tensions may develop between the older residents and the newcomers.

Key words

counter-urbanisation
urban–rural migration
suburbanised villages

Management of counter-urbanisation

The green belt

As cities and towns grow they spread out into the surrounding countryside. One way decision-makers try to stop the countryside being taken over by urban growth is by creating a **green belt** around cities. London was the first city to have a green belt (1947) but the map shows that many others have followed suit.

The aims of a green belt are:
- to stop **urban sprawl**
- to prevent towns and cities merging into each other
- to protect the countryside
- to encourage development within the town, not around it

Designation of land as green belt does not mean it will never be built on. If a developer can prove that building is needed, green belt land can be used. This happened when the M25 motorway was built round London and when Manchester airport was given permission to expand.

Key
Urban area
Green belt

Newcastle

0   km   100

N

Bradford   York
Manchester   Leeds
Liverpool
Sheffield
Stoke-on-Trent
Nottingham
Coventry
Birmingham
Cambridge
Gloucester
Oxford
London
Bristol
Southampton

Green belts in England and Wales (2006)

Today the enormous need for new houses means that some areas of green belt land have been built on and others are threatened. Britain needs about 3 million new homes by 2020, mostly in southeast England. There are not enough **brownfield sites** (land which has been built on before) for all the extra houses, so there is pressure to build on green belt.

It is cheaper for developers to build on greenfield sites on the edge of towns than to redevelop brownfield sites in older parts of a town or city

New housing in the green belt

There is a demand for large detached and semi-detached houses in pleasant surroundings. These houses can fetch a high price

Permission has been given for 10,000 new homes in green belt land west of the town of Stevenage in Hertfordshire, despite huge opposition from environmental groups

Modern houses have excellent facilities inside and a garage and small garden outside

Key words

green belt
urban sprawl
brownfield site

Migration within England

About 10% of England's population move house each year. Most people stay within the same region but a significant number of people move to other parts of the country. Coastal areas have the biggest gain from internal migration and cities are the biggest losers.

The **southwest** is also a popular area to move to. Migrants may be moving to work in cities such as Bristol, or moving to live in the countryside or by the coast. Many people choose to move to the south coast when they retire.

The **southeast**, excluding London, is the most popular region to move to because this is where the greatest number of jobs is available, either by travelling into London or in the region itself. Some migrants move from other parts of the UK and some move from London to the southeast and commute to work.

■ Areas gaining population
■ Areas losing population

The **northeast**, the **northwest** and the **West Midlands** have all experienced population decline in the last decade. This is because these areas have seen increases in unemployment as manufacturing industries have closed and because there are not so many job opportunities here, especially in the recent recession.

East Anglia is an area which has more people moving into than out of the area.

London is the area with the highest out-migration. Although there are lots of jobs available in London it is also an expensive and busy place in which to live. Many people decide to move to the surrounding areas to live, especially if they have a family, and retired people may also choose to live elsewhere.

Why do people move?

■ Most people move to find work, so areas with high unemployment tend to lose population and areas which have more jobs or better paid jobs gain population.

■ Older people often choose to move away from large cities when they retire perhaps to the countryside or to a coastal area. This is an example of counter-urbanisation.

■ Many young adults move away from home either to go to university or to find a job.

■ As people earn more money they may choose to move out of towns and cities and live in rural areas and commute to work. This is an example of counter-urbanisation.

Consequences of population change

■ Areas which are gaining people, such as southeast England, southwest England and East Anglia, may become overcrowded and expensive and services such as healthcare and education may come under pressure.

■ Areas which are losing population, such as London and the north, may become unattractive to new businesses and provide fewer opportunities for the people who live there.

Managing population change in the UK

The UK government tries to make sure that migration is not causing problems within the country and that some areas do not become much more prosperous than others. The government supports the least prosperous areas by giving grants and loans to businesses and investing in infrastructure, particularly transport. This can help bring jobs to the area.

The UK government is also moving some of its own workers away from London and the southeast to other centres such as Manchester, Leeds and Newcastle.

Test yourself

1 **Cross out the incorrect word(s) from those in italic in the following.**
   (a) Sparsely populated areas are *too cold, too flat, too dry,* and *too mountainous* to live in.
   (b) Densely populated areas are *industrial, infertile plateaus* or *rich in natural resources.*
   (c) Overpopulated areas have *too few/too many people* for the resources available.
   (d) Birth rate is the number of live babies born per *100/1,000/10,000* of the population each *month/year.*

2 **Complete the sentences.**
   (a) In Stage 2 (early expanding) of the demographic transition model the death rate is ...
   (b) Two features of Stage 3 (late expanding stage) of the demographic transition model are ...
   (c) Five push factors causing people to migrate are ...

> **Exam tip**
>
> Command words such as 'describe' and 'explain' are very important (explain is the same as asking for reasons). Notice that the final part of this question asks for a named place you have studied. You must start your answer by naming a country or you will score very few marks.

Examination question

(a) Look at the diagram which shows population pyramids for two countries, A and B. One is an LEDC and the other an MEDC.
   (i) Which of the two pyramids is for an LEDC? *(1 mark)*
   (ii) Explain how changes in the death rate may affect the size of the population. *(2 marks)*
   (iii) What planning problems might result in a country with a population structure similar to that of country A? *(4 marks)*
   (vi) Why has the rate of population growth in some countries declined? *(4 marks)*

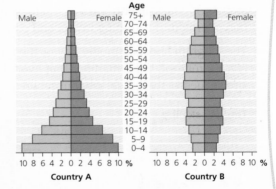

Case study: strategies to influence natural population change within a country

Foundation tier:
(b) Name one country you have studied where the government has tried to influence natural population change.

   Describe the ways the government tried to influence population change. Explain some of the problems caused by this strategy. *(8 marks)*

Higher tier:
(b) Name a country you have studied where strategies have been introduced to influence natural population change.

   Describe the strategies which were used and the extent to which these have been successful. *(8 marks)*

Land use in MEDC cities

Land use in a city falls into these categories:

Residential — land used for housing. This takes up the largest area. Most cities have many different residential areas.

Open space — land used for parks and playgrounds, usually spread throughout the city; derelict or unused land is also open space and is usually found in older, more central, areas.

Commercial — land used for shops, offices, banks and other businesses. Although commercial areas such as shopping centres are found in many parts of the city, the biggest concentration is in the central business district (CBD).

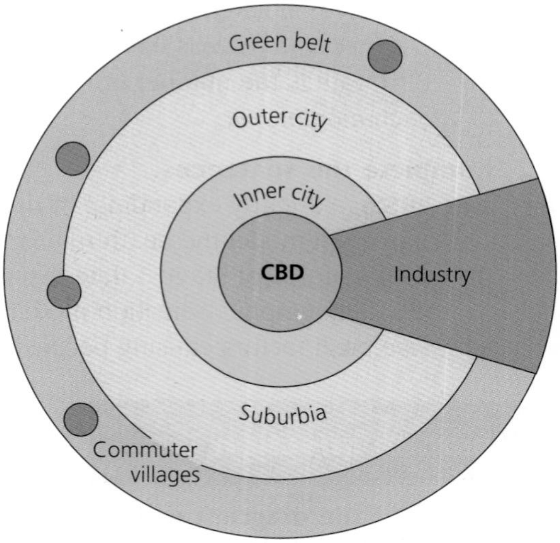

The model city — land use in an MEDC city

Green belt — many cities are protected by a green belt where new building is restricted. The aim is to reduce **urban sprawl**. If cities are allowed to grow outwards without restriction they may merge into adjoining towns and cities forming a **conurbation**.

The diagram shows a model city or the typical layout of land use in an MEDC city. This model is a useful tool, but remember that every town or city is unique. Towns and cities may show similarities to the model city but nowhere will be arranged just like the model.

The city skyline

Key words

land use	central business district (CBD)	leisure facilities	doughnut effect
green belt	commercial functions	quarter	congestion
urban sprawl		accessible	
conurbation		redevelopment	

The central business district (CBD)

The CBD is at the heart of the city. It is dominated by high-rise buildings occupied by shops, offices, banks and other **commercial functions**. There are often **leisure facilities** such as theatres, cinemas, night clubs, restaurants and pubs, and these may be clustered together in a **quarter**. All these functions group together in the CBD because it is the most **accessible** part of the city. People from all over the city and beyond can reach it easily. This pushes up the value of the land and also explains why there are so many high-rise buildings. The high cost of rents in the CBD means that some land uses are not found here, such as housing, industry and large areas of open space.

Changing land use in the city centre

The CBD is always changing. Some businesses move away as others move in. City councils are keen to keep businesses and shops in the city centre as this provides jobs and generates money. Some changes are on a small scale but in some city centres large areas have been **redeveloped** as old buildings are demolished to make way for new, modern developments.

If too many businesses move away from the CBD, for example into out-of-town business parks or shopping malls, then there is a danger of leaving a 'dead heart' in the city centre, sometimes called the **doughnut effect**.

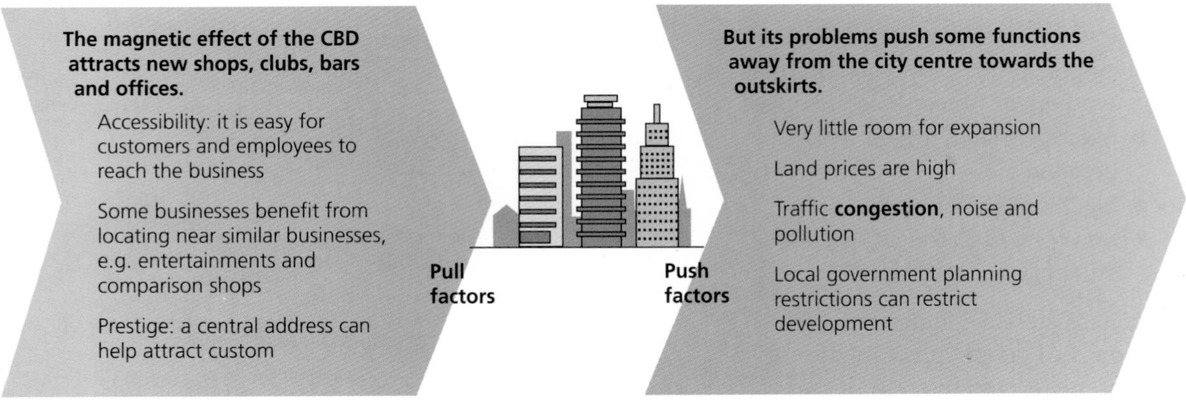
The magnetic effect of the CBD attracts new shops, clubs, bars and offices.
Accessibility: it is easy for customers and employees to reach the business
Some businesses benefit from locating near similar businesses, e.g. entertainments and comparison shops
Prestige: a central address can help attract custom
Pull factors
Push factors
But its problems push some functions away from the city centre towards the outskirts.
Very little room for expansion
Land prices are high
Traffic congestion, noise and pollution
Local government planning restrictions can restrict development

Push and pull factors affecting functions in the CBD

Changes include:

- new buildings
- shopping malls
- public open spaces
- conversion of old buildings for a new use
- pedestrianised areas and one-way streets
- new types of transport such as trams (e.g. Manchester and Sheffield)

The inner city

Most **inner-city areas** lie close to the CBD and to industrial parts of the city. The inner city is an area of poor-quality, often **nineteenth-century terraced housing**. In the 1960s some of these old terraces were cleared as part of **urban redevelopment schemes** and replaced with council estates including **high-rise flats**. Today many inner-city estates are run down even though they are only about 40 years old. Some 1960s flats have been knocked down

High-rise flats built in the 1960s
Terraced houses built over 100 years ago. Some have become run down
High-density housing and little open space
Parking is often a problem

Inner-city housing

Why people leave inner-city areas

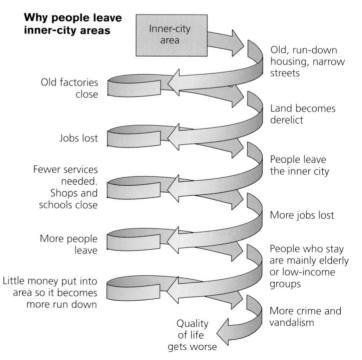

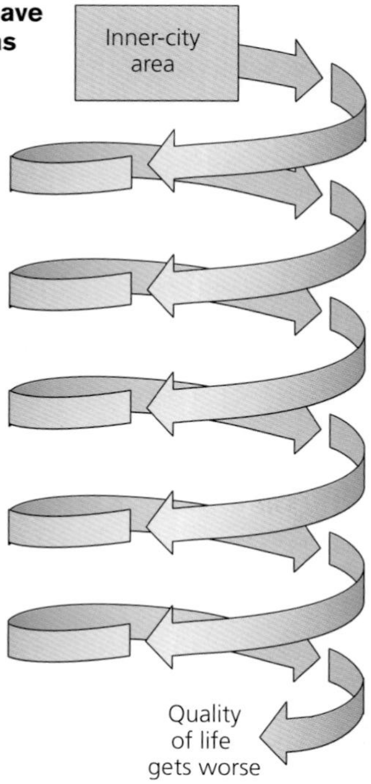

Inner-city area
Old, run-down housing, narrow streets
Old factories close
Land becomes derelict
Jobs lost
People leave the inner city
Fewer services needed. Shops and schools close
More jobs lost
More people leave
People who stay are mainly elderly or low-income groups
Little money put into area so it becomes more run down
More crime and vandalism
Quality of life gets worse

People of the inner city

Many different types of people live in inner-city areas but certain groups are found in quite high numbers:

- low-income families such as single parents, low-paid manual workers or unemployed people
- older people living on a state pension
- black or Asian families
- newcomers to the city such as students and migrants

Census statistics indicate which groups of people predominate in each ward (voting area) of the city. We call these the **socio-economic characteristics** of the area.

People living in inner-city areas may have a low **quality of life** because of low incomes, poor-quality housing and inadequate social facilities.

Changes in the inner city

Many inner-city areas have undergone huge changes during the last few years, and quality of life for people has begun to improve. Local councils can only afford small-scale schemes but large schemes can be implemented with government or European funding. Once initial changes have been made it is often possible to attract private investors into the area which helps it to regenerate. It is important that any changes improve the area for the long term, in other words they are **sustainable**.

Industries move out

In the past, when factories and businesses were located in the inner city, local people could find work within reach of their homes. Today many old factories have closed down because manufacturing has declined. Others have moved away to the outskirts of the city where land is cheaper, there is more space to expand and access to motorways is easier. The narrow streets of the inner city are not attractive to new businesses. Large-scale redevelopment may be necessary to improve both the environment and employment prospects in the area.

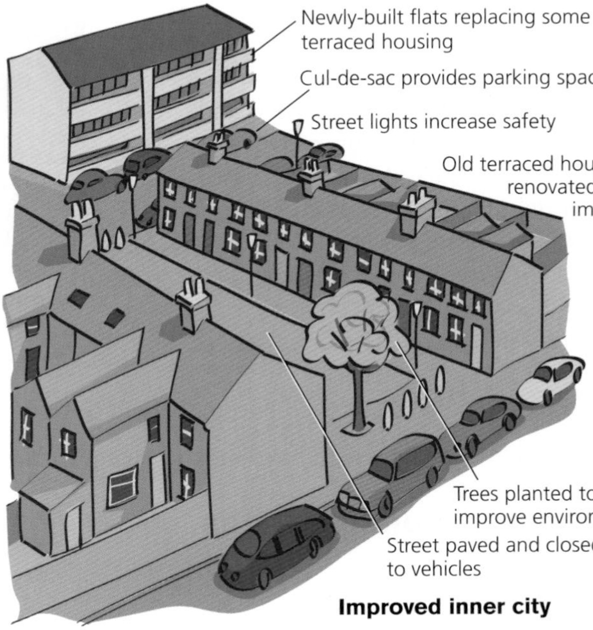
Newly-built flats replacing some terraced housing
Cul-de-sac provides parking space
Street lights increase safety
Old terraced houses renovated and improved
Trees planted to improve environment
Street paved and closed to vehicles

Improved inner city

Waterfront redevelopment

Many cities have central waterfront areas along a coastline, river or a canal. In cities such as London, Manchester and Liverpool once-busy docks had become derelict and have now been redeveloped, as has land alongside canals in cities such as Birmingham and Leeds. In these areas substantial investment has improved the **infrastructure** of the area by clearing derelict land, improving roads and access and constructing new buildings. This was followed by private investment as property developers moved in building houses, offices and retail outlets. In this way the areas have been brought back to life and become prosperous again. New residents and businesses move in and the areas are often attractive to visitors as well. This creates jobs and helps the local economy as well as improving the appearance of the area.

Key words

- inner city
- nineteenth-century terraced housing
- urban redevelopment schemes
- high-rise flats
- socio-economic characteristics
- quality of life
- sustainable
- infrastructure

The outer city

Since the beginning of the twentieth century, cities have spread outwards rapidly. Transport in the form of trams, then buses and much later cars meant that people could live further away from their place of work. It led to the growth of the outer city or **suburbs**.

Suburbia

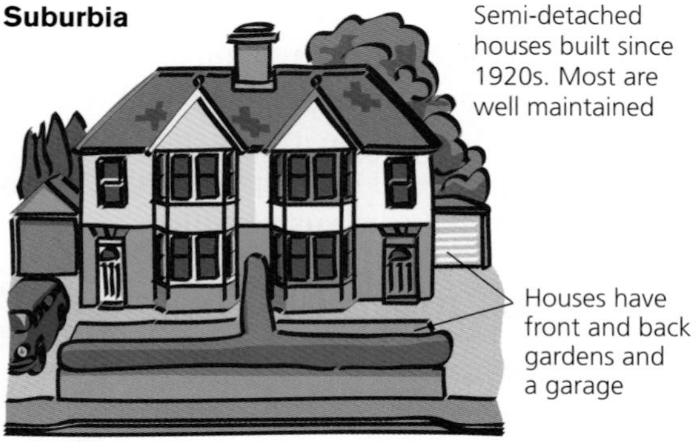
Semi-detached houses built since 1920s. Most are well maintained
Houses have front and back gardens and a garage

There are many different types of houses in the suburbs, including detached and semi-detached houses, bungalows and flats. The houses are built at lower densities than in the inner city, more of them have gardens and there are more parks and other areas of public open space. There is a huge variety of housing because:

- different people need, and can afford, different sorts of houses
- building styles have changed enormously during the last 100 years

Exam tip

If you are writing about an urban area in an exam, use a named area from a town or city you have studied.

Areas of the outer city which have mostly owner-occupied houses are called **suburbia**.

Outer-city council estates

There may also be large **council estates** in the outer city, often built on **greenfield sites** during the 1960s and 1970s when there was huge demand for housing. The estates usually have a mixture of high-rise and low-rise houses together with a shopping centre, school and open space. In some cities these are popular and successful but in some they have become rundown housing areas with the residents feeling isolated from the central parts of the city.

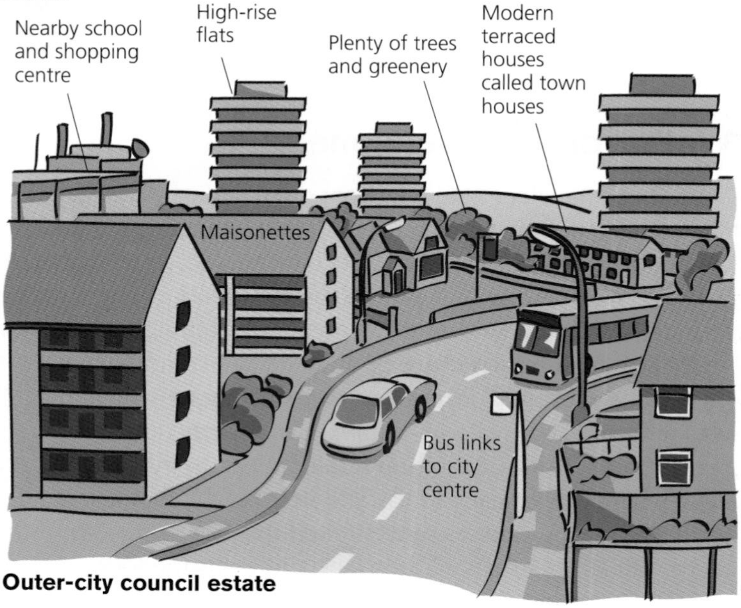
Nearby school and shopping centre
High-rise flats
Plenty of trees and greenery
Modern terraced houses called town houses
Maisonettes
Bus links to city centre

Outer-city council estate

The rural–urban fringe

Land on the outskirts of the town changes slowly from a built-up urban area to countryside. This area, where town and countryside merge, is known as the **rural–urban fringe**. It usually has both urban and rural land uses.

Changes to the rural–urban fringe

Land in the rural–urban fringe is often targeted for new developments such as housing estates, shopping malls, office parks or waste disposal facilities. Land on the outskirts is often cheaper and more readily available than land within the urban area. Main roads and motorways can make this area accessible and greenfield sites are preferred by developers.

Decision makers must try to balance the need for sustainability with the demand for new developments. Sometimes developments are allowed to proceed but sometimes plans for using greenfield sites will be rejected. Controversial or large-scale proposals, for example for new roads, housing estates or new airport runways, may be referred to a public enquiry before a decision is made.

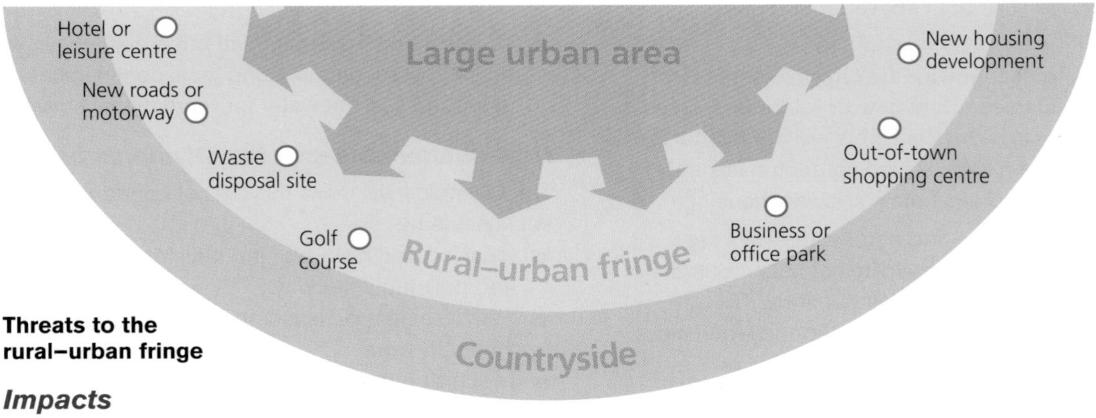

Threats to the rural–urban fringe

Impacts

These new developments provide good opportunities for local businesses and provide services and facilities for local people including employment opportunities and leisure facilities.

There are problems with developments in the rural–urban fringe.

- Countryside is lost as the town spreads outwards. This could be land used for both recreation or for farming.
- Traffic increases, causing more pollution and congestion.
- Inner city areas and **brownfield sites** in the city may remain undeveloped as it is easier and cheaper to develop greenfield sites in the rural–urban fringe.

Some cities have **green belt** land around them to protect the rural–urban fringe from development (see page 55).

Key words

- suburbs
- suburbia
- council estate
- greenfield site
- rural-urban fringe
- brownfield site
- green belt

Case study: *urban change to illustrate social, economic and environmental planning and its sustainability*

East London

Hosting the 2012 Olympic Games in London has led to a major urban regeneration project in the Stratford area of east London. Before the successful bid this was a rundown area which had suffered from de-industrialisation and had large expanses of industrial waste land. The successful Olympic bid gave the area a kick start for redevelopment and now it is at the centre of the plans for the 2012 Games. Much of the Olympic Park site was polluted brownfield land which is now being cleaned; a huge range of facilities are being built here including the Olympic Stadium, Aquatics Centre, velopark, two indoor arenas and the Olympic Village where the athletes and team officials will live during the Games. Public transport is being improved by extending and upgrading rail links both above ground and underground.

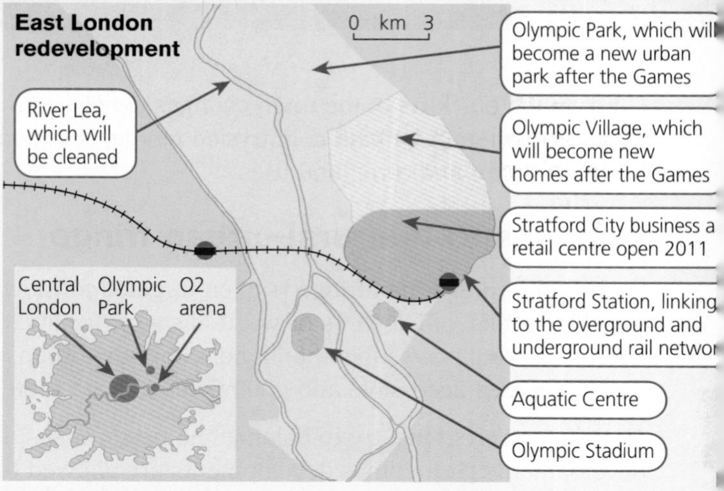
East London redevelopment
0 km 3
River Lea, which will be cleaned
Olympic Park, which will become a new urban park after the Games
Olympic Village, which will become new homes after the Games
Stratford City business a retail centre open 2011
Stratford Station, linking to the overground and underground rail networ
Aquatic Centre
Olympic Stadium
Central London
Olympic Park
O2 arena

In addition to changes for the Olympic and Paralympic Games a vast new development called Stratford City is being built. This will open in 2011, and consists of a large shopping centre, leisure facilities, office space, hotels and new housing.

The nearby River Lea and other waterways are being cleaned and new wildlife habitats created including a new wetland. This will attract wildlife, provide recreation space for local people and also help to manage flood risk in the area.

After the games

It is hoped the Olympics will benefit the area in the future, as this is an important way to make the development sustainable. After the Games:

- part of the Olympic Park will be used to create the largest urban park in Europe
- the Olympic Stadium and five sports venues will be used by local people and also for specialist sports training. Others, such as the hockey stadium, will be dismantled and used elsewhere
- the Olympic village will provide permanent homes for 9,000 people. Many will be 'affordable' homes which will be offered mainly to key workers, for example, teachers and nurses
- about 12,000 permanent jobs will be created
- there will be better public transport and better roads
- access to the River Lea and other waterways will be increased, encouraging local communities to enjoy and use the water for recreational activities

How sustainable can the Olympics be?

The Olympics will make this part of London more sustainable by

- cleaning and redeveloping old industrial land and polluted waterways
- establishing a parkland and open space in this inner city area
- revitalising the waterways for recreation and wildlife
- generating permanent jobs
- providing new homes
- encouraging local people to walk, cycle and use public transport instead of cars
- making the area attractive which will encourage more businesses to be established and in turn lead to more jobs

However, to make way for the Olympic Park 400 people had to be re-housed from the site of the Olympic Village and 350 companies had to be relocated from the site of the main stadia, causing loss of some jobs when these companies moved.

Test yourself

1 **Copy and complete this table.**

2 **Describe how self help schemes help to tackle problems caused by urbanisation in LEDC cities.**

3 **Explain why the rural–urban fringe is under threat from developers.**

	Causes	Consequences
Urbanisation	(1) (2)	▪ Overcrowded housing ▪ ▪ ▪
Counter-urbanisation	▪ ▪ ▪	▪ Expansion of rural settlements ▪ ▪ ▪

Examination question

Population of London falls despite rise in migrants

Almost 1.8 million people have moved to London from abroad over the last decade, new figures revealed today, however nearly 2 million people moved out of the capital to other parts of Britain between 1998 and 2007. This 2 million was equivalent to more than one-quarter (26%) of the city's population and is part of a national trend which has seen people leaving the cities for the green fields, clean air and lower crime rates of the countryside.

(a) Read the newspaper article above and suggest two reasons why people are moving away from inner-city areas. *(2 marks)*

(b) In what ways might the quality of life of people living in a small town or village be better than that of people living in inner-city areas? *(3 marks)*

Case study: *a sustainable urban development scheme*

Foundation tier:

(c) (i) Name and locate an urban development scheme you have studied.

(ii) Describe how this area has changed or will change in the future

(iii) Explain the ways in which the changes are sustainable. *(8 marks)*

Exam tip

Command words such as 'describe' and 'explain' are very important (explain is the same as asking for reasons). Notice that the final part of this question asks for a case study (named place) you have studied. You must start your answer by naming an urban development scheme or you will score very few marks.

Higher tier:

(c) Name and locate an example of a sustainable urban development scheme. Describe the main features of the scheme and explain how it is sustainable. *(8 marks)*

Sustainable cities

About half of the world's population live in urban areas. Most cities are **unsustainable**, in other words they consume raw materials, energy and water and produce harmful waste, and as cities grow they take up more land and damage the environment. The planet cannot support this and so ways must be found to reduce the impact cities are having and make them more **sustainable**. Making urban areas sustainable means changing people's lifestyles and the way they think about the environment, and changing how cities are planned.

Ways to make cities more sustainable

- Build zero-carbon homes and offices
- Improve public transport so people do not use cars as much
- Use brownfield sites for new developments so derelict land is re-used instead of building on greenfield sites
- Use local food supplies instead of transporting food around the world
- Reduce waste by re-using products such as bottles and plastic containers and recycling glass, paper and textiles
- Involve local people in decision making
- Provide more open space and greenery to improve quality of life
- Make sure people live closer to their work so they can walk or cycle
- Use renewable energy sources such as wind, water or solar power

Zero-carbon living

About 25% of the UK's carbon emissions come from homes. It is possible to reduce this by improving insulation and installing double glazing to prevent heat loss and by encouraging people to reduce the amount of energy they use, for example by switching to low-energy light bulbs, turning off standby switches and turning down heating by a few degrees. **Carbon-neutral homes** are those where carbon emissions do not add to the net amount of carbon dioxide in the atmosphere.

Key words

- unsustainable
- sustainable city
- carbon neutral homes
- zero-carbon homes

The government has announced that all new homes in the UK should be **zero-carbon** by 2016, this means not releasing any carbon dioxide at all into the atmosphere. This is a big challenge as most electricity is still generated by burning fossil fuels so new homes will need to make their own energy, for example by having solar panels and using boilers fuelled by biomass.

The world's first zero-carbon city is Masdar City, which is being constructed in Abu Dhabi. Construction of this desert city began in February 2008, and it is designed to be powered entirely with renewable energy, including solar power.

Eco-towns

In July 2009 the government announced that four **eco-towns** would be built, in England. Each will have 4,000 to 5,000 homes and building will start by 2016. It is hoped another six eco-towns will be developed later.

All houses will be zero-carbon and generate their own energy from renewable sources such as the sun or wind; residents will be able to sell surplus energy to the national grid.

The towns will have smart meters to track energy use, community heat sources and charging points for electric cars.

About 40% of the towns will be parks, playgrounds and gardens. There will be efficient public transport, cycle routes and footpaths, and shops and a primary school within easy walk of every single home.

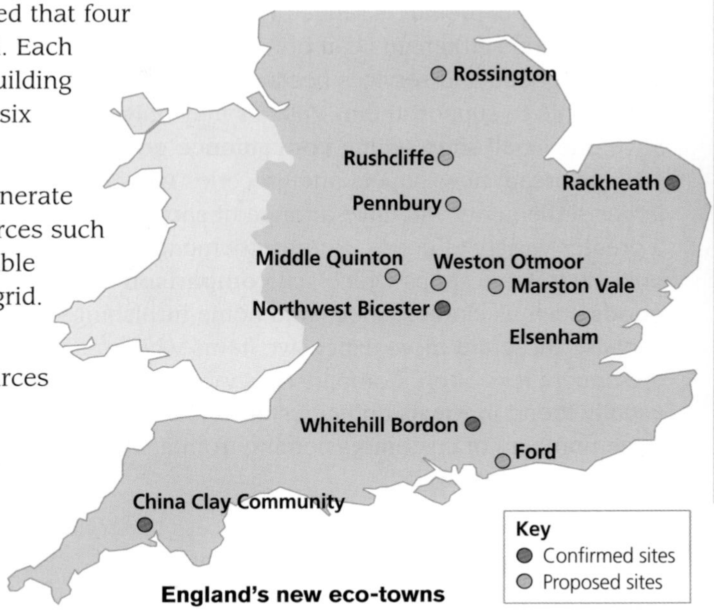

England's new eco-towns

Some people think eco-towns are a good idea because

- the towns are planned to be sustainable
- they will provide thousands of new houses and many will be 'affordable' homes for lower income families
- they will help to tackle climate change by reducing emissions of greenhouse gases

Some people think eco-towns are a bad idea because

- they will be built on greenfield sites using up valuable countryside and green belt land
- too few houses will be built to meet demand
- they will generate more traffic which will generate more carbon emissions

Bed-ZED, Wallington, South London

In 2002 the Beddington Zero Energy Development or 'Bed-ZED' was opened. This is an experiment in zero-carbon living in which 100 households live in apartments designed to save as much energy as possible. For example, they use wind turbines and biomass boilers to generate power, have well-insulated buildings and collect and use rainwater. People who live here use public transport or bikes although they also have some shared cars for use when needed. The community also buys locally produced food and recycles its waste.

This community has shown that it can make large savings on heating, electricity and water consumption and reduce its ecological footprint.

Key words

eco-town

Retail services in settlements

All settlements provide a range of services for local people. Rural settlements can only provide a limited range of goods and services because there are too few people to support them. Villages may only have one small shop selling **convenience goods** such as bread, newspapers and groceries whereas larger settlements will have a range of shops selling a greater variety of goods. Rural settlements are unlikely to have shops which sell **comparison goods** such as clothes, shoes and home furnishings because these are more expensive items which are bought less often. Comparison shops are usually found in towns and cities which have large numbers of customers coming from a wide **catchment area**.

Some groups of people have better access to services than others and this can affect quality of life.

Distribution of shopping centres in a large town

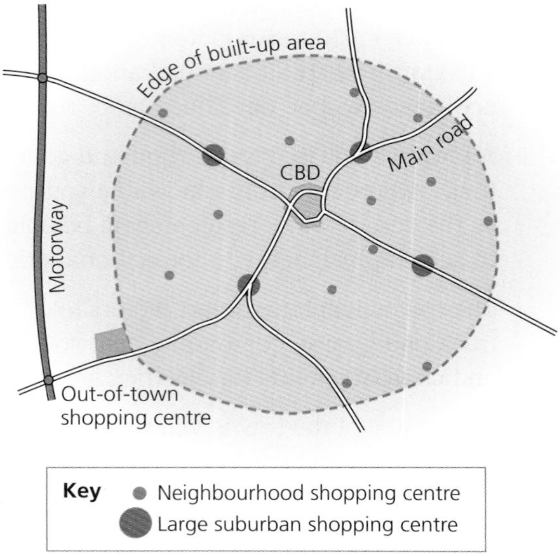

Shopping centres in a town

The diagram shows that most towns have:

One major shopping centre in the CBD which contains the largest shops such as department stores and chain stores. These need lots of customers to support them (they have a **high threshold**) and so they locate in the most accessible part of the city. The shops may sell **high-order goods**, such as expensive clothes, electrical equipment and furniture.

Several large suburban shopping centres, usually located towards the centre of the city at major road junctions. These locations provide good access from surrounding residential areas. As well as supermarkets and other food shops there will be banks, estate agents, charity shops and hardware stores.

Many neighbourhood shopping centres with anywhere between six and 60 shops. They will mostly sell **low-order goods** and services such as food, newspapers, fish and chips and hairdressing. These shops have a **low threshold** and rely on frequent visits from people living nearby.

Corner shops used to form an important part of the shopping **hierarchy** but many of them have closed as they are unable to compete with large supermarkets or other businesses which sell cheaply and stay open for long hours.

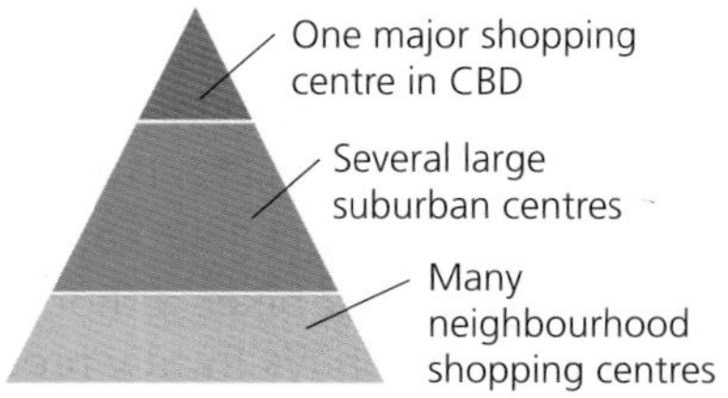

Hierarchy of shopping centres in a town

Changes in retail services

Today the distribution of shops and shopping centres is more complex than in the past. Improvements to roads and motorways mean shopping centres can be developed away from the main built-up area and the congestion of city centres. There are also many different ways to purchase goods, particularly by going online, and this has had an impact on retail services.

Out-of-town shopping centres

In recent years changes to people's lifestyle have affected shopping patterns:

- more people have more money to spend (disposable income)
- many more families have a car, giving them access to new shopping centres
- electrical appliances such as freezers and microwave ovens mean people buy food for several days or weeks in advance and no longer buy fresh food every day
- more women work and have less time to shop every day than in the past

Some traditional shopping centres have suffered as their customers prefer to shop elsewhere.

Today many cities have:

- several **superstores** on the outskirts
- out-of-town **retail parks** with large electrical stores, carpet warehouses and computer shops
- large **out-of-town shopping centres** such as the Metro Centre in Gateshead, the Trafford Centre in Manchester and Meadowhall in Sheffield

These types of retail developments are generally located on the outskirts of a city where land values are lower and there is more space available to build and expand. It is important for them to be close to main roads or motorways so there is easy access for both their customers and delivery vehicles, and away from the congestion of the city centre.

E-commerce or internet shopping (or e-tailing)

During the last decade more and more goods and services have been available online and this is becoming an important way in which people shop. In some ways this has given retailers an exciting new opportunity and many companies, large and small, have websites which facilitate online shopping.

Huge new businesses which only trade online have also been established, for example Amazon, and these are in competition with traditional shops.

Internet shopping has also put pressure on high street shops as people may prefer to stay at home to use the internet to shop instead of spending their money in shopping centres. Currently about 20 pence in every pound is spent online but the proportion is growing, which is putting pressure on retailers that do not have an online presence.

Key words

convenience goods
comparison goods
catchment area
high threshold
high-order goods
low-order goods
low threshold
hierarchy
superstores
retail parks
out-of-town shopping centres

Case study: *illustrating how retail service provision changes over time*

Changing retail services in Sheffield

Large cities like Sheffield have a huge range of shops and retail services but recent changes mean that there is now more choice and more variety for customers.

New shopping areas have become established because of improved transport within the city. Many more people today own their own cars but improved public transport, especially the super tram, means that access to some parts of the city has improved.

Changing consumer demands and market forces have meant that some shops have closed and that others have opened.

In some cases, completely new retail areas have become established.

Many large supermarkets are found in Sheffield, particularly in the southwest which is the most affluent part of the city. The range of goods and services has changed; in addition to food and drink the largest stores sell toys, health and beauty products, clothing and shoes. Many have their own pharmacy and cafe. Their aim is to sell 'everything under one roof'. Most have embraced internet shopping by providing a delivery service for products ordered online.

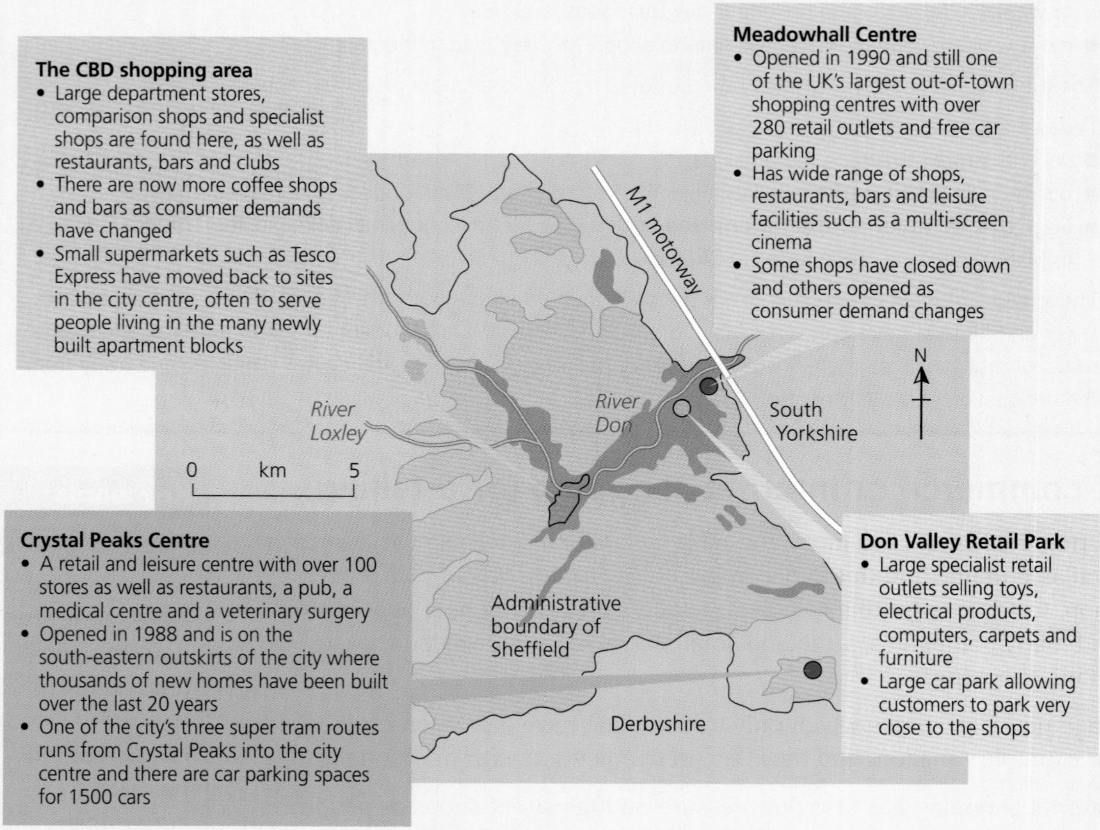
The CBD shopping area
• Large department stores, comparison shops and specialist shops are found here, as well as restaurants, bars and clubs
• There are now more coffee shops and bars as consumer demands have changed
• Small supermarkets such as Tesco Express have moved back to sites in the city centre, often to serve people living in the many newly built apartment blocks
Meadowhall Centre
• Opened in 1990 and still one of the UK's largest out-of-town shopping centres with over 280 retail outlets and free car parking
• Has wide range of shops, restaurants, bars and leisure facilities such as a multi-screen cinema
• Some shops have closed down and others opened as consumer demand changes
M1 motorway
N
River Loxley
River Don
South Yorkshire
0 km 5
Crystal Peaks Centre
• A retail and leisure centre with over 100 stores as well as restaurants, a pub, a medical centre and a veterinary surgery
• Opened in 1988 and is on the south-eastern outskirts of the city where thousands of new homes have been built over the last 20 years
• One of the city's three super tram routes runs from Crystal Peaks into the city centre and there are car parking spaces for 1500 cars
Administrative boundary of Sheffield
Derbyshire
Don Valley Retail Park
• Large specialist retail outlets selling toys, electrical products, computers, carpets and furniture
• Large car park allowing customers to park very close to the shops

Test yourself

1 Look at the diagrams of an unsustainable city and a sustainable city.

(a) Identify the differences between the two diagrams.

An unustainable city

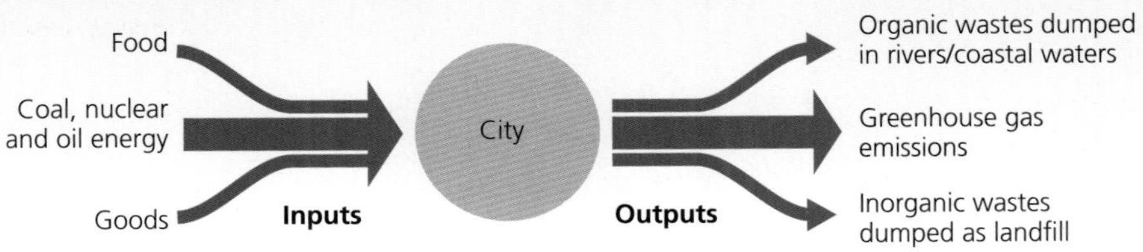

A sustainable city

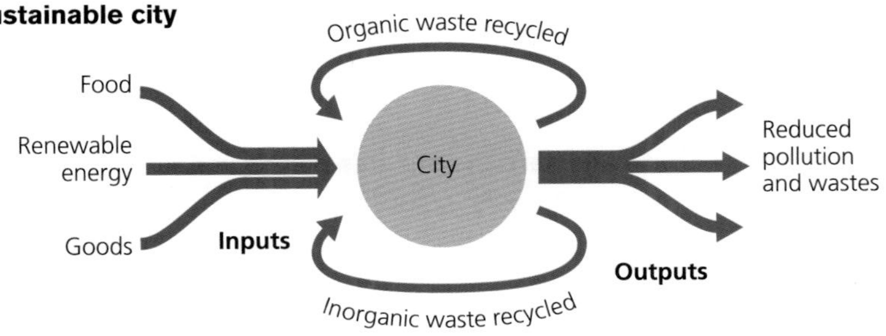

(b) In what ways will the sustainable city be less damaging to the local environment?

(c) In what ways will the sustainable city be less damaging to the global environment?

2 **(a)** Give reasons why the government wants to develop eco-towns in the UK.

(b) Name two groups of people who might be opposed to eco-town development and give reasons why they have these views.

3 Define these terms: high-order goods, comparison shops, catchment area, out-of-town shopping centre, e-commerce.

Examination question

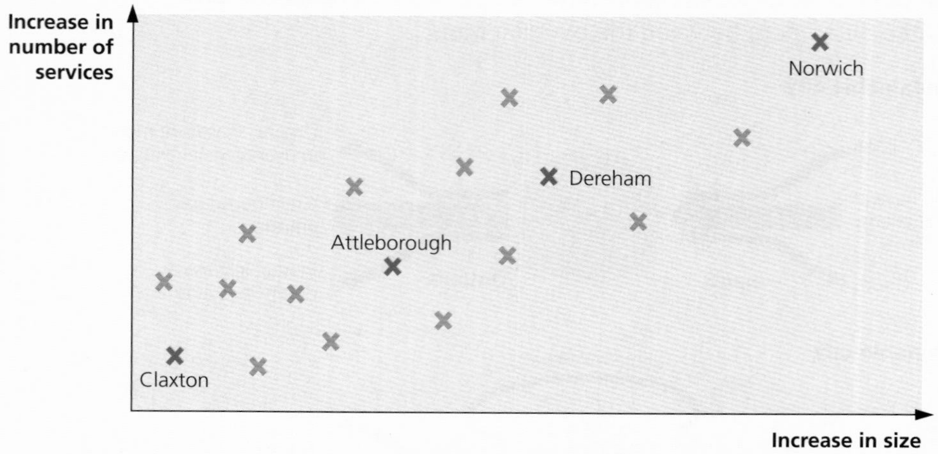

Size and services of some places in East Anglia

(a) **Look at the diagram, which shows the size and services of some places in East Anglia.**

- **(i)** **What is meant by services?** *(1 mark)*
- **(ii)** **Which place named on the graph has the most services?** *(1 mark)*
- **(iii)** **What does the graph show about these places in East Anglia?** *(2 marks)*

(b) **(i)** **What is meant by counter-urbanisation?** *(1 mark)*

- **(ii)** **Give four reasons why people move to the countryside from the cities.** *(4 marks)*

Exam tip

If there is a graph in the question make sure you know what the axes represent, and look for general trends and unusual points.

Case study: *changes in retail services over time*

Foundation tier:

(c) **Name and locate an area you have studied where the types of shops and services have changed over time.**

Describe how shops and services in this area have changed.

Give three reasons why these changes have happened. *(8 marks)*

Higher tier:

(c) **Name and locate a place where retail services have changed in recent years.**

Describe how retail services have changed and give reasons for the changes. *(8 marks)*

Theme 3

Natural hazards

Plate tectonics

The Earth's crust is made up of pieces like a jigsaw, called **tectonic plates**.

The plates 'drift' or move because the heat in the rock below the crust sets up convection movements (rather like the movement you see in a pan of soup when it is heated).

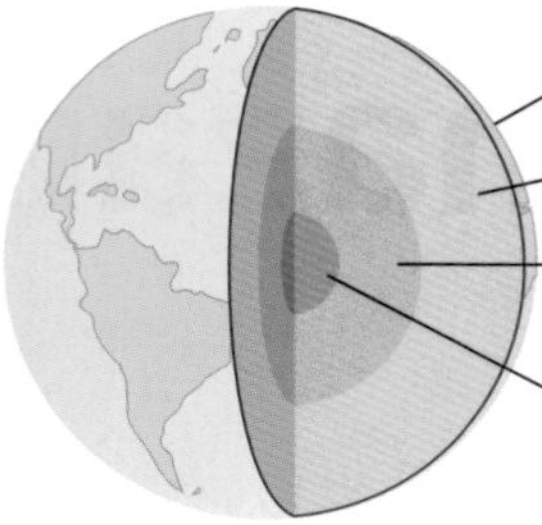

The Earth

The tectonic plates

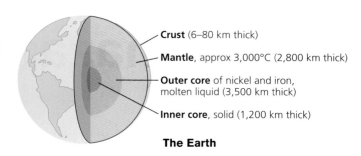

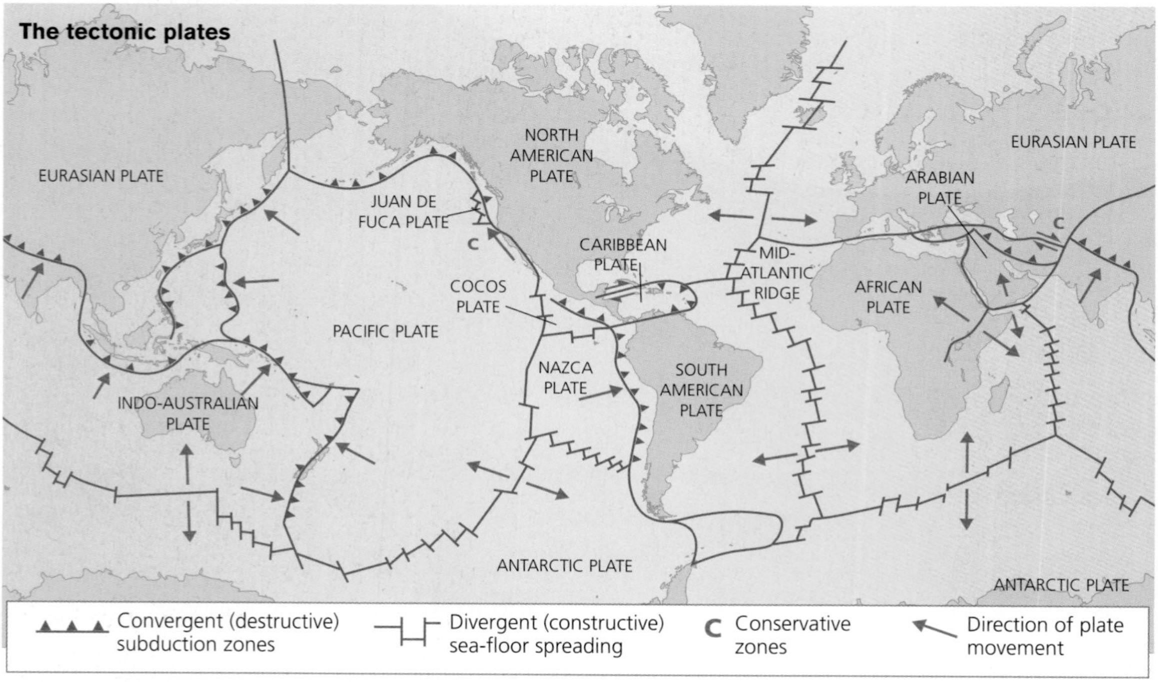

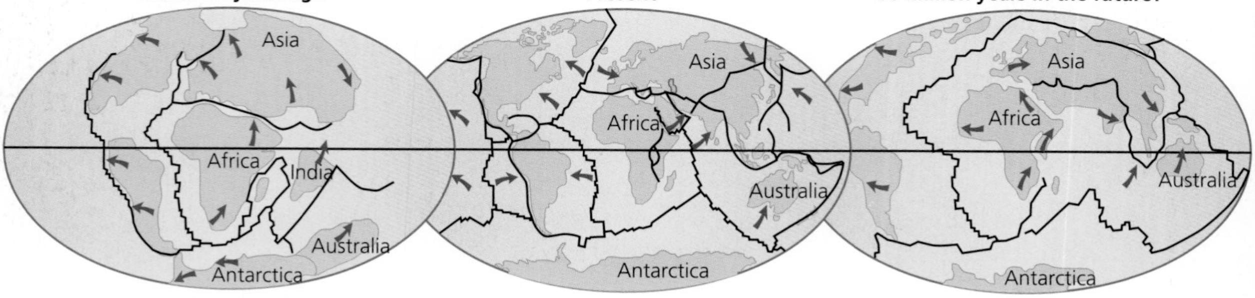

How the plates have moved

The movement of tectonic plates

The plates are moving all the time. At the margins (edges) they either move apart, together or slide past each other.

Plate movement

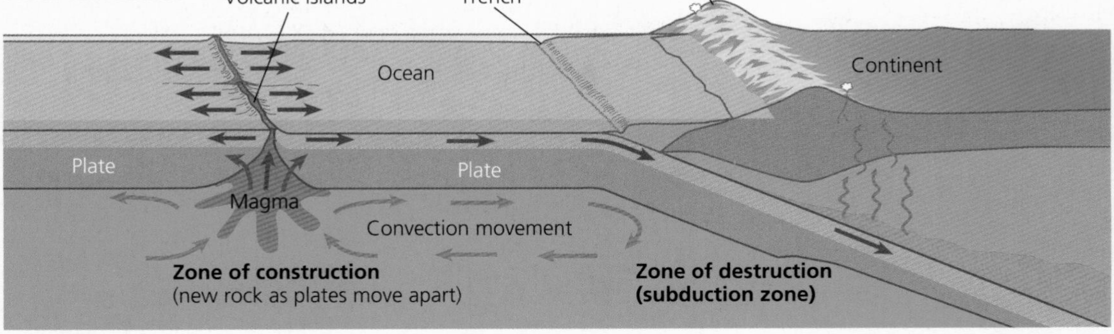

Types of plate margin

Plates meet at plate margins. There are four different types.

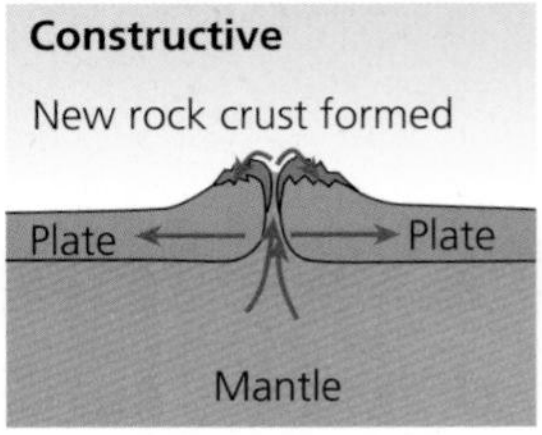

At a **constructive plate margin** the plates are moving apart. Molten rocks from the mantle below spread out and harden, forming a ridge of new rock. Here the Earth's crust is under great tension and cracks known as faults appear. It is this movement which causes earthquakes. Volcanoes occur as the molten material wells up, sometimes creating volcanic islands. For example, the mid-Atlantic ridge and Iceland were formed in this way.

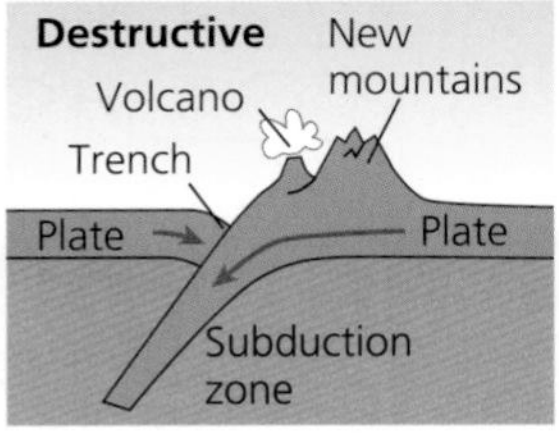

A **destructive plate margin** occurs where one plate slides beneath another as they collide. The bottom plate crumples, creating new mountains. Pressure builds up as friction stops the plates sliding past each other, when they do move this is usually with a 'jerk', which releases large amounts of energy as earthquakes. Molten rock is released as the lower plate pushes further into the Earth, which reaches the surface as a volcano, for example, the Andes Mountains and volcanoes of South America.

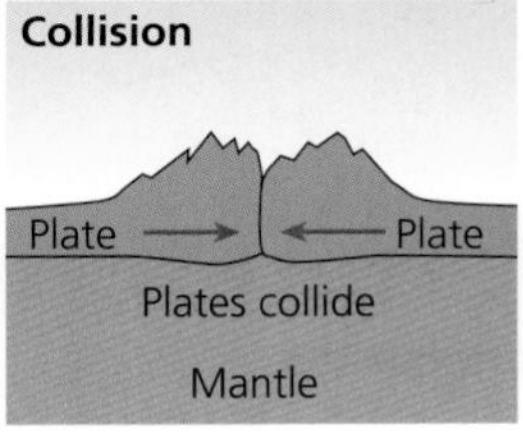

A **collision plate margin** is where two plates collide and are crushed against each other. They are pushed upwards, forming new mountains. Earthquakes are common but there are no volcanoes. For example, the Himalayas were caused by the collision of the Indo-Australian and the Eurasian plates.

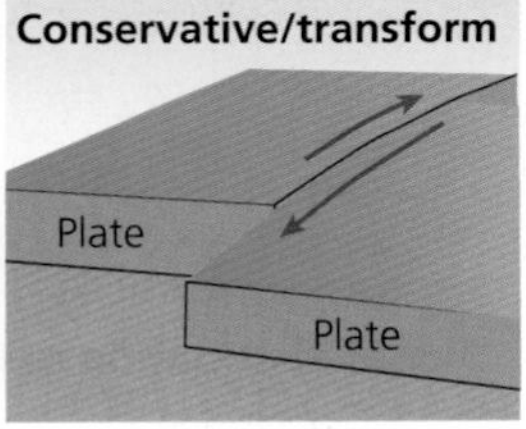

At a **conservative** or **transform plate margin** the plates slide past each other. A fault line can usually be seen on the Earth's surface. Pressure builds up until the plates move with a 'jerk', causing earthquakes. For example, the San Andreas fault in California caused an earthquake which hit San Francisco.

Key words

- inner core
- crust
- mantle
- outer core
- zone of construction
- zone of destruction/subduction zone
- conservative or transform plate margin
- constructive plate margin
- destructive plate margin
- collision plate margin
- tectonic plates

Earthquakes

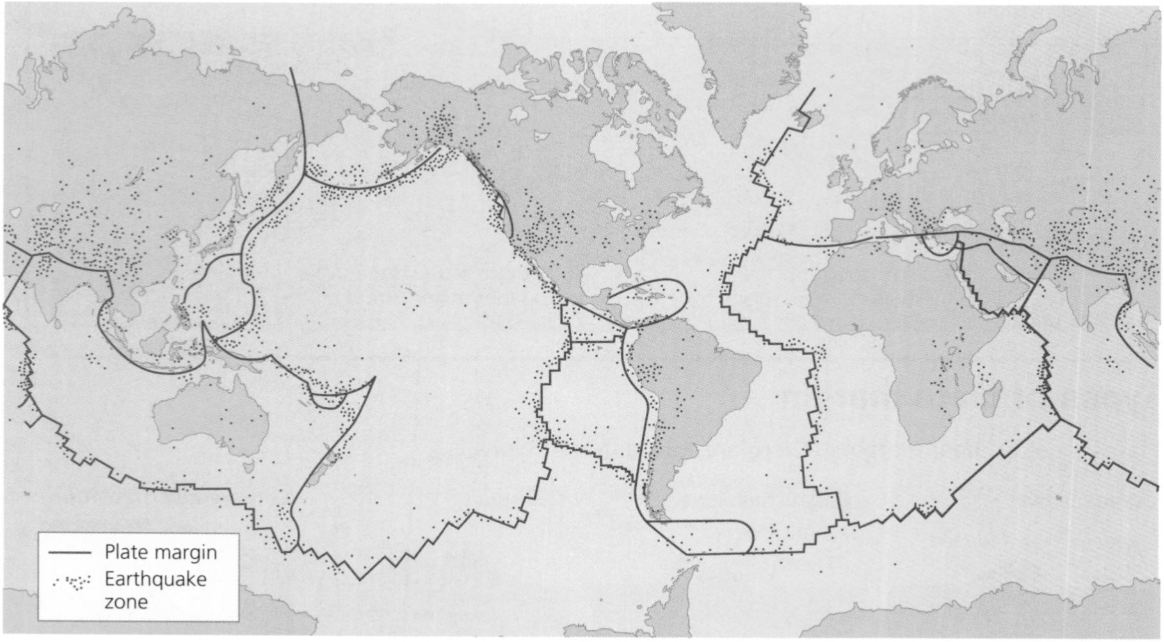

Plate margin
Earthquake zone

Global distribution of earthquakes

Earthquakes and volcanoes are examples of natural hazards that occur on the plate margins. The major earthquake zones are around the Pacific Ocean and in mountainous regions such as the Himalayas, central China and Iran. Not all earthquakes occur on land; many happen beneath the sea where they can cause enormous tidal waves known as **tsunami**, which can devastate coastal areas.

Earthquakes are usually the result of plate movement that is *not* smooth. The strain builds up along a fault line between two plates until they move suddenly, causing earthquakes.

The point where the earthquake starts below the Earth's surface is known as the **focus** or **hypocentre**. The point directly above the focus on the Earth's surface is known as the **epicentre**.

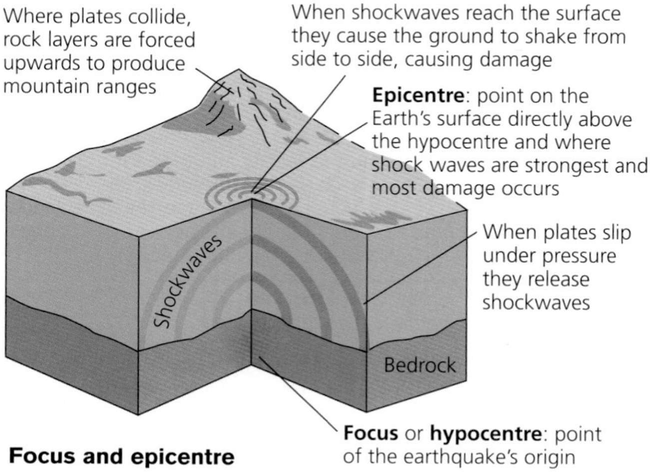

Where plates collide, rock layers are forced upwards to produce mountain ranges
When shockwaves reach the surface they cause the ground to shake from side to side, causing damage
Epicentre: point on the Earth's surface directly above the hypocentre and where shock waves are strongest and most damage occurs
Shockwaves
When plates slip under pressure they release shockwaves
Bedrock
Focus or **hypocentre**: point of the earthquake's origin

Focus and epicentre

Earthquakes: the effects

Primary (immediate) effects

- Ground shaking — buildings shake and collapse, causing loss of life.
- Cracks appear in the ground — gas pipes are broken causing fires to break out.
- Communication links are broken, e.g. transport and telephone.
- Power supplies are disrupted.
- Water pipes are broken and supplies disrupted.

Secondary effects

- Shaking ground may cause sands to turn to a liquid state causing yet more buildings to collapse and disruption to communications.
- Aftershocks, which may occur days after the initial earthquake, cause further building collapse and transport disruption.
- Landslides bury roads and buildings and may dam rivers creating dangerous 'quake lakes'.
- Earthquakes under the ocean may cause tsunamis.
- Outbreaks of disease in the population may occur because of disruption to water and medical supplies.

Measuring earthquakes

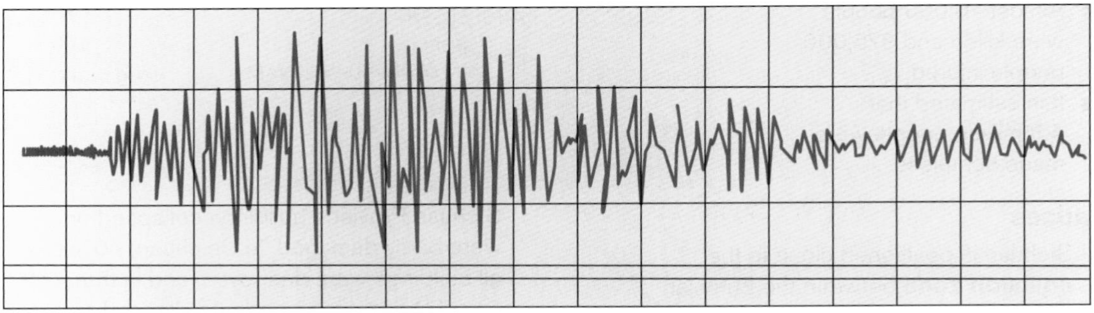

A seismograph of an earthquake

The magnitude of an earthquake is recorded by an instrument called a **seismometer**. It measures the height of the shock waves on the **Richter scale**. Each point on the scale is ten times greater than the one below. This means that an earthquake measuring 7 is ten times more powerful than one with a score of 6. The seismometer records the earthquake vibrations with a pen on a sensitive arm, marking zigzag lines on a drum of paper.

The **Mercalli scale**, which has been largely replaced by the Richter scale, measures the strength of an earthquake by its observed effects on buildings. A score of 1 on the Mercalli scale would only be detectable by seismographs and there would be no damage at all, while 12 would be catastrophic, with no buildings left standing.

The strongest earthquake ever recorded was in Chile in 1960. It measured 9.5 on the Richter scale.

The earthquake off the coast of Sumatra in the Indian Ocean which led to the 2004 tsunami measured 9.3 on the Richter scale.

Key words

tsunami
focus
hypocentre
epicentre
seismometer
Richter scale
Mercalli scale

Case study: *a tectonic hazard event in an LEDC*

The 2008 China earthquake

Fact file

- The earthquake in Sichuan Province struck on 12 May 2008.
- It measured 7.9 on the Richter scale.
- The epicentre was Wenchuan County, 92 km northwest of the city of Chengdu.
- The tremor was felt in Beijing (1,500 km away) and Shanghai (1,700 km away).
- Almost 70,000 people were killed and 375,000 people injured.
- It is estimated that 4.8 million people were made homeless.

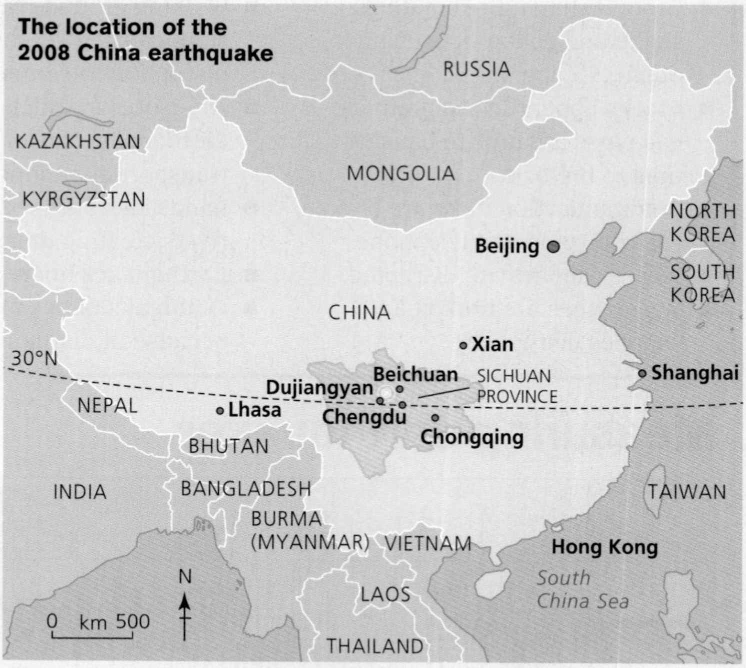

Causes

- Sichuan is positioned close to the **collision zone** between the Eurasian plate and the Indian plate.
- The Indian plate is moving towards the Eurasian plate along the collision zone.
- These plates became temporarily stuck, causing a build up of pressure.
- The pressure built up until it was released with a sudden jolt along a major **fault**.
- This slippage along the major fault caused the earthquake.

The primary (immediate) effects

- The ground shaking caused cracks to appear in buildings in Chengdu, although no buildings collapsed.
- As soon as the tremors were felt, buildings across China were evacuated.
- In many towns and villages throughout Sichuan Province buildings collapsed or were badly damaged. In Beichuan, 80% of all buildings were destroyed, and in the city of Dujiangyan a school collapsed. Fewer than 60 of the 900 children survived. Transport was disrupted: all the roads in Sichuan Province were severely damaged, the railway line was distorted and the airport was closed.
- Water mains were broken so no water was able to reach houses.
- Power supplies were cut as power lines and hydroelectric power stations were destroyed. The lack of power and mobile phone congestion devastated wireless communications.

The secondary effects

- Landslides buried roads and buildings, making the work of relief teams more difficult.

- Heavy rains increased instability on steep slopes and landslides dammed rivers causing water to pond back, forming lakes. There was a risk that the dams holding the water of these 'quake' lakes would burst, causing flooding and more loss of life. Channels were built using explosives to drain the lakes.
- Concern grew about the risk of disease outbreaks because of the lack of water, food and medical supplies.

Relief efforts

- The Chinese authorities responded immediately — helicopters, troops and medical teams were sent to the area.
- Many governments and voluntary organisations responded rapidly — the Red Cross and Red Crescent sent tents, medical supplies water and food.
- Transport carrying aid was hampered by landslides and many villages waited several days for aid to reach them.

Impact of the earthquake

- The devastating impact was partially due to the severity of the earthquake (7.9 on the Richter scale) and the high density of population in the earthquake area.
- China is an **LEDC** (a less economically developed country) and although China has vast economic and human resources, LEDCs are not as well prepared for such catastrophes as **MEDCs** (more economically developed countries).
- Many buildings were old and the poor design and poor building materials and methods contributed to the loss of life as houses collapsed.

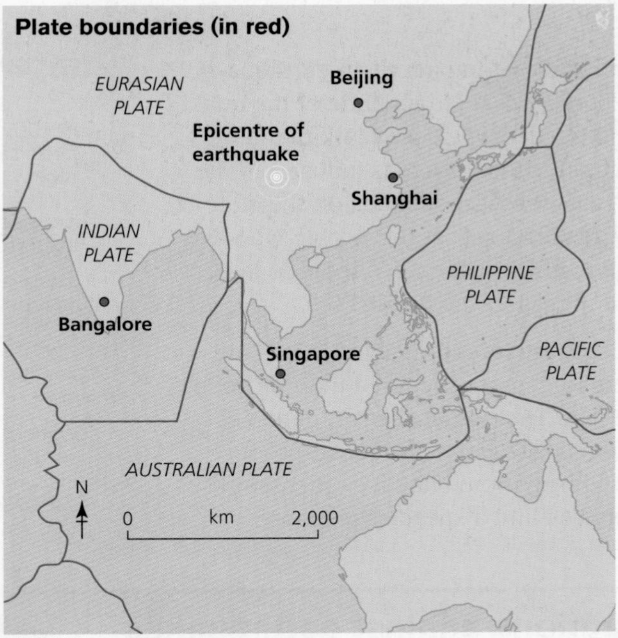

- Buildings were constructed on known fault lines.
- A large number of schoolchildren died as poorly constructed schools collapsed. Because of China's 'one child' policy many parents lost their only child. The 'one child' restriction was quickly relaxed in Sichuan Province.
- In this agricultural area irrigation systems, greenhouses and agricultural buildings such as barns were destroyed.
- Many animals died, both wild and domesticated. It is estimated that 1 million pigs died.
- However, short-term planning was effective and a massive relief effort was mounted rapidly.
- It will take more than 3 years to rebuild the damaged areas at a cost of US$50 billion.

Key words

collision zone
fault
LEDC
MEDC

The impact of earthquakes

The effect or impact of an earthquake depends on the magnitude of the quake and the density of population and human activity such as industry in the area. But it also depends on the state of preparedness of the people who live there which will enable them to lessen the impact of the event.

People and authorities in MEDCs are generally more prepared than those in LEDCs. Their greater wealth enables them to develop earthquake-proof buildings, more effective emergency services and a speedier response.

Impact of earthquakes in LEDCs

Why the impacts of earthquakes in LEDCs are often worse than in MEDCs

- If LEDCs require foreign aid to cope with the aftermath, it takes time to arrive
- Lack of services so there are limited medical supplies and hospitals to deal with injured
- Communications are poor: access roads are poorly maintained and in remote areas people may not have radios
- The people are not prepared and do not know what to do in the emergency
- Housing is built using traditional methods and is not earthquake-proof
- Without adequate water the spread of disease is a constant danger
- There are few experts available to give advice and assess the situation
- There are few plans for dealing with the emergency, which means delays in bringing emergency services to the area

Precautions against earthquakes

Individuals

- prepare an emergency pack including water, food, blankets, first aid kit, radio and torch
- during and after the earthquake, shelter under a table or bed; avoid stairways
- turn off gas, water and electricity
- after the earthquake move to open ground

Authorities

- monitor the hazard so people can be warned
- have emergency supplies ready
- make plans for shelter, food and water supplies, for emergency services, fire brigade, police, ambulance and hospital services
- plan to broadcast information for people affected

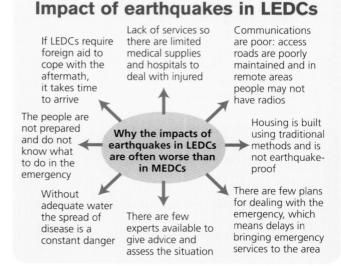

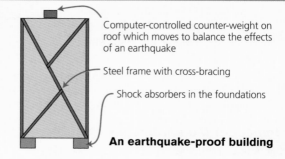

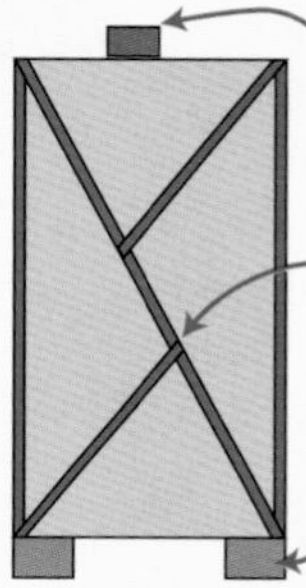

An earthquake-proof building

Long-term planning

Long-term planning is more likely to be sustainable:

- ensure road and rail communications are built to reduce the effect of earthquakes
- ensure all new buildings are earthquake proof
- through education and advertising ensure that all people know what to do in the event of an earthquake

Predicting earthquakes

Although a great deal is known about where earthquakes are likely to occur, it is impossible to predict exactly the days or months when an earthquake will happen in any specific location. The only really successful prediction was in China in 1975 when an evacuation warning was issued the day before the earthquake struck. The prediction was based on changes in the height of the land, levels of underground water, animal behaviour and, most tellingly, a sudden increase in the number of 'foreshocks'.

The United States Geological Survey attempted to predict activity on the San Andreas Fault in California, one of the most-studied earthquake areas in the world. Using the most sophisticated instruments available they suggested that an earthquake would occur between 1988 and 1992; an earthquake occurred in 2004.

Why people live in earthquake zones

Despite the known dangers people still live in earthquake zones.

For example, people live in California — a known earthquake hazard area — because the warm, sunny climate is favourable for growing crops using irrigation water. It is a coastal area with a large port and city, San Francisco, and a range of hi-tech industries. These factors provide a desirable lifestyle and people are willing to live there despite the earthquake hazard.

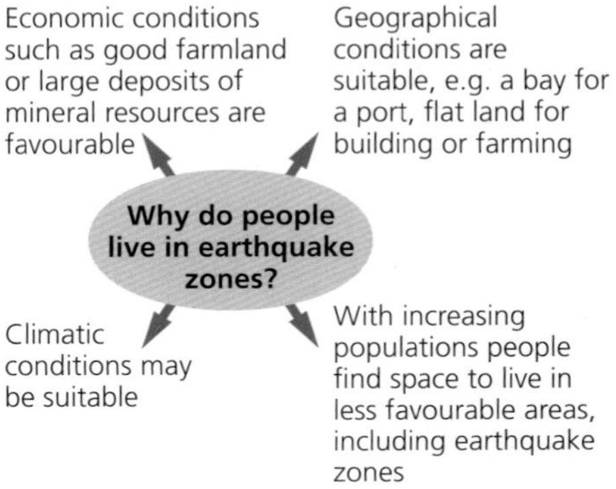
Why do people live in earthquake zones?
Economic conditions such as good farmland or large deposits of mineral resources are favourable
Geographical conditions are suitable, e.g. a bay for a port, flat land for building or farming
Climatic conditions may be suitable
With increasing populations people find space to live in less favourable areas, including earthquake zones

Test yourself

1 Label the diagrams correctly using the list of words to the right.

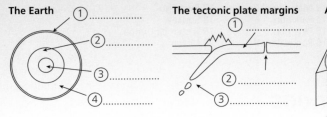

The Earth

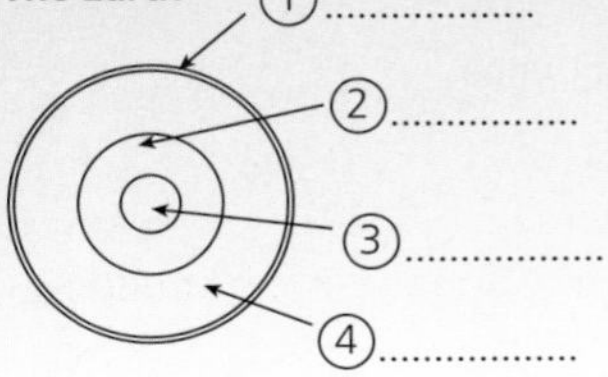
1 ....
2 ....
3 ....
4 ....

The tectonic plate margins

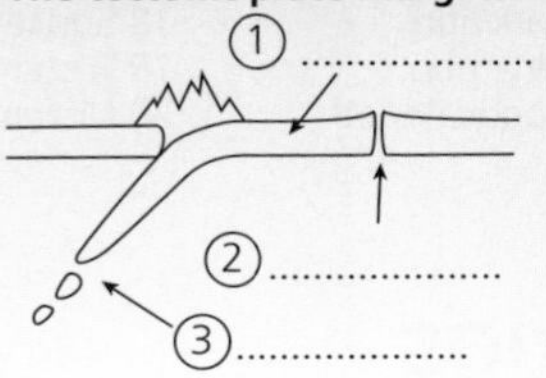
1 ....
2 ....
3 ....

An earthquake

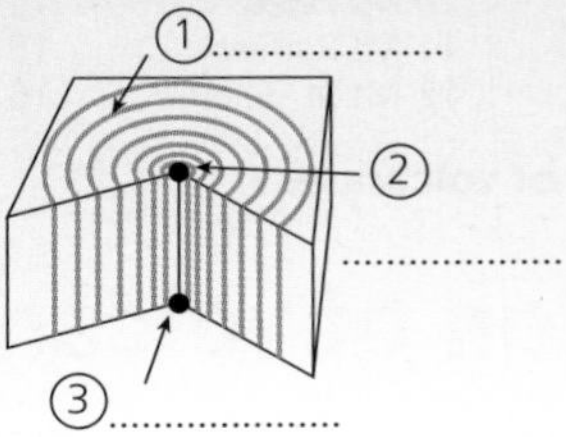
1 ....
2 ....
3 ....

epicentre
inner core
primary
zone of construction
seismometer
shock waves
Sichuan
mantle
fault
crust
landslides
Mercalli scale
zone of destruction
focus
outer core
plate

2 Fill in the blanks with the correct word from the list to the right.

Earthquakes are measured using a ....

The .... measures the strength of an earthquake by its observed effects on buildings.

The ground shaking and cracking is a .... effect of earthquake action.

The 2008 earthquake in China occurred in .... Province.

One secondary effect of the earthquake in China in 2009 was ....

A crack or weakness in the Earth's crust is known as a ....

Volcanoes

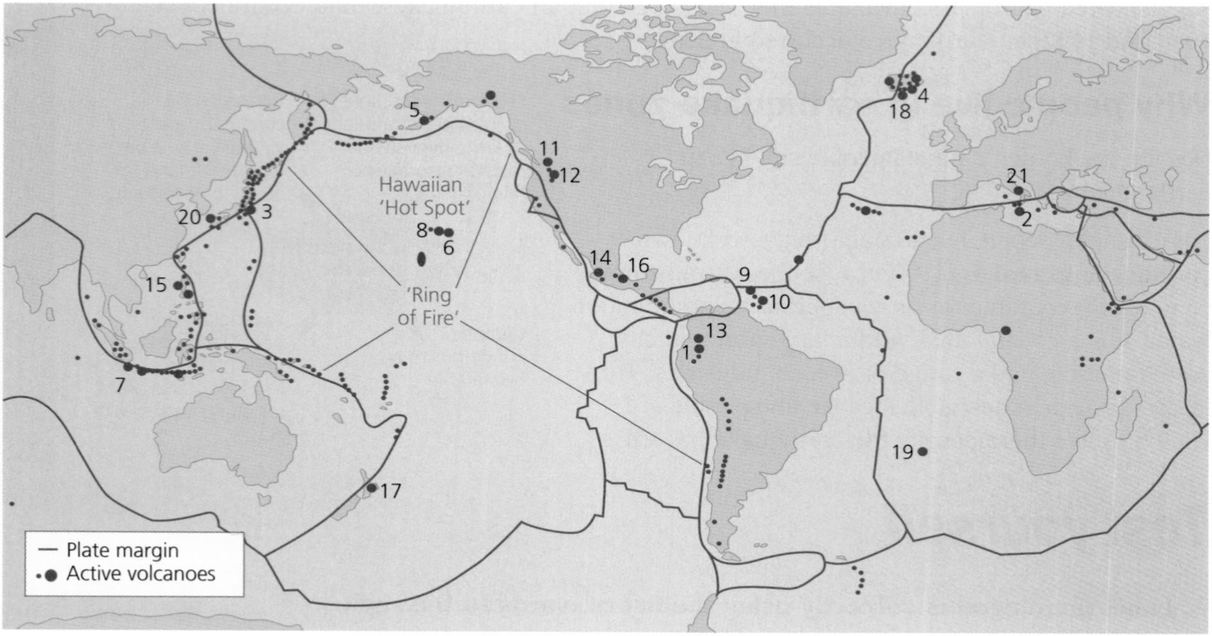

1 Cotopaxi
2 Etna
3 Fujiyama
4 Heimaey
5 Katmai
6 Kilauea
7 Krakatoa
8 Mauna Loa
9 Soufrière Hills
10 Mt Pelée
11 Mt Rainier
12 Mt St Helens
13 Nevado del Ruiz
14 Parícutin
15 Pinatubo
16 Popocatépetl
17 Ruapehu
18 Surtsey
19 Tristan da Cunha
20 Unzen
21 Vesuvius

Global distribution of volcanoes

Distribution of volcanoes

The majority of volcanoes occur at the tectonic plate margins and their distribution is similar to that of earthquakes. Particularly noticeable is the concentration around the Pacific Ocean mirroring the plate margins of the Pacific Plate and known as the 'Ring of Fire'.

Why volcanoes occur

Molten rock known as **magma** can escape when pressure builds up below the Earth's surface. When the magma reaches the surface it is known as **lava**.

Volcanoes may erupt very explosively at destructive plate margins, releasing enormous amounts of lava, ash and steam. Because such volcanoes are so violent, throwing rock debris long distances at high speed, they cause devastation in the blast zone. Ash, which is also thrown out of such volcanoes often to great heights, can blanket large areas surrounding the volcano, sometimes burying towns. Such volcanoes, called **strato-volcanoes**, are of two main types, known as **composite volcanoes** and **dome volcanoes**.

Sometimes the lava is more liquid and flows more gently to the surface through cracks in the crust at constructive pate margins. Such flows of lava can move very quickly and devour everything in their path. Such volcanoes are known as **shield volcanoes**.

When a plate is over a particularly hot part of the mantle, known as a **hot spot**, it may be pierced and magma may escape to the surface forming a shield volcano. Such volcanoes may form away from the plate margin.

Types of volcano

Active volcanoes are those which have erupted recently and are likely to erupt again.

Dormant volcanoes are those which have not erupted for a long time but may erupt again.

Extinct volcanoes are unlikely to ever erupt again.

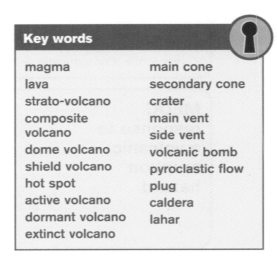

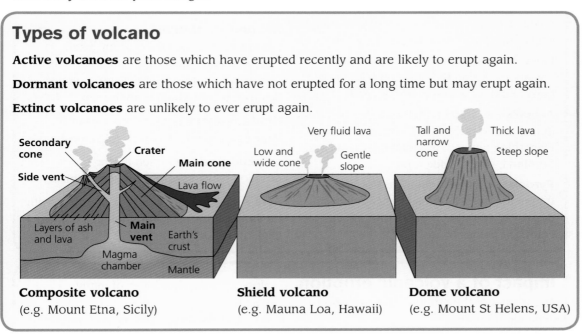

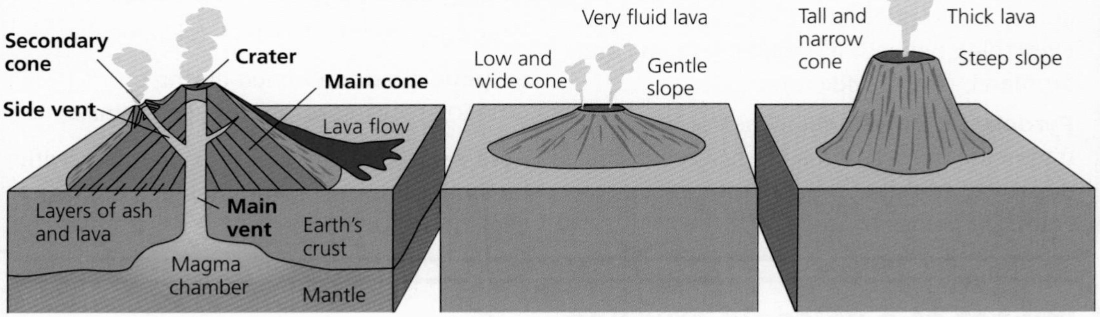

Composite volcano
(e.g. Mount Etna, Sicily)

Shield volcano
(e.g. Mauna Loa, Hawaii)

Dome volcano
(e.g. Mount St Helens, USA)

Volcano facts

- When a volcano erupts violently it throws ash and **volcanic bombs**, which are pieces of rock, into the air.
- Hot gases, ash and steam can form **pyroclastic flows** which move very fast and can cause tremendous damage.
- When a volcano has been dormant for some time the solidified magma in the vent acts as a **plug**. When the volcano erupts the plug is blown out, often blowing off the top of the cone and leaving a very large crater known as a **caldera**.
- Mud flows, called **lahars**, are formed when hot ash melts snow and ice or falls into rivers. They move very fast and are very destructive.
- There are more than 600 active volcanoes in the world.

Key words

magma
lava
strato-volcano
composite volcano
dome volcano
shield volcano
hot spot
active volcano
dormant volcano
extinct volcano
main cone
secondary cone
crater
main vent
side vent
volcanic bomb
pyroclastic flow
plug
caldera
lahar

Volcano hazards

The effects of a volcanic eruption

Primary effects

Volcanic blasts Explosive blasts from volcanic eruptions carry rock debris at high speed and cause widespread devastation.

Lava flows Molten rock flowing from volcanoes during the eruption will crush, burn and bury everything in its path.

Ashfalls Ash thrown high into the air during an eruption falls to Earth covering everything beneath it, including towns and farmland, over a wide area.

Pyroclastic flows Avalanches of hot ash, fragments of rock and superheated gas destroy everything in their path and can cause great loss of life.

Earthquakes May be triggered by violent volcanic eruptions.

Secondary effects

Lahars Mudflows, which occur when heavy rainfall causes ash and other debris to be washed down the volcano's slopes, destroying everything in their path.

Landslides Violent eruptions trigger landslides on the slopes of volcanoes.

Sulphur dioxide Emitted during an eruption causing respiratory problems.

Acid rain When sulphur dioxide mixes with rainwater it forms acid rain which damages vegetation.

Impact of a volcanic eruption

The impact of the eruption of a volcano is dependent upon:

- the violence of the eruption
- the density of population in the vicinity of the volcano. This is often high because ash and lava that has been broken down by the elements is usually very fertile. Consequently people live close to or on the slopes of many volcanoes
- how well the eruption has been predicted and whether effective warning systems are in place
- how well-prepared the people who live near the volcano are and how quickly the authorities warn them and rescue services act

MEDC response to a volcanic eruption hazard

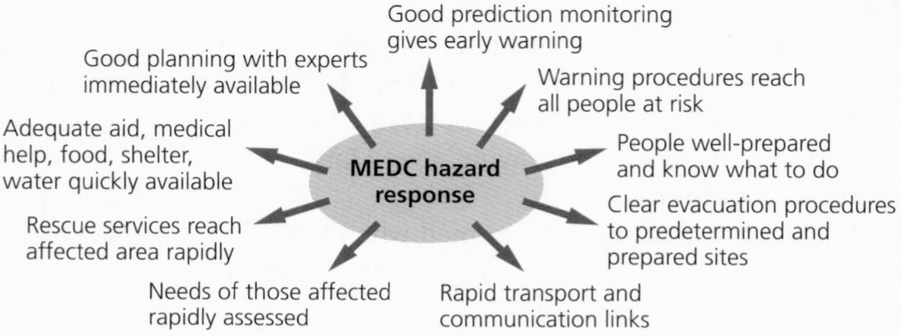
MEDC hazard response
Good prediction monitoring gives early warning
Warning procedures reach all people at risk
People well-prepared and know what to do
Clear evacuation procedures to predetermined and prepared sites
Rapid transport and communication links
Needs of those affected rapidly assessed
Rescue services reach affected area rapidly
Adequate aid, medical help, food, shelter, water quickly available
Good planning with experts immediately available

Reducing the impact of a volcanic eruption

Prediction Effective monitoring and prediction enable authorities to warn people living nearby to evacuate the area. A number of indicators warn of impending eruption:

- the frequency of earth tremors
- the movement of magma inside the volcano can be measured using gravity measurements as the Earth bulges under the pressure
- the build up of magma causes ground temperatures to rise
- emissions of gases such as sulphur dioxide increase

Evacuation When it is judged that an eruption is imminent, effective warning and evacuation procedures can dramatically reduce the loss of human life.

Mapping The path of old lava flows, mudflows and lahars can be mapped to show areas of greatest risk.

Engineering Impact can be lessened by direct intervention, minor lava flows can be diverted by bulldozing walls to turn the flow away from villages. In Iceland seawater has been sprayed onto lava flows to cause them to solidify and stop their flow.

Why people live near a volcano:

- The weathered lava and ash is a very rich and fertile soil for farming.
- Visitors come to see volcanoes and so people can find jobs in the tourist industry.
- Geothermal energy can be harnessed and used to produce electricity.
- Rocks from the volcanoes are rich in minerals and can be mined.

Key words

volcanic blast
lava flow
ashfall
pyroclastic flow
earthquake
lahar
landslide
sulphur dioxide
acid rain

Test yourself

1 Label the diagram correctly using words from the list on the right.

2 Match the correct word on the right to each description:

- A volcano which has not erupted for a long time but may erupt again.
- Hot gases, ash and steam which move very fast and can cause tremendous damage.
- A mud flow formed when hot ash melts snow and ice or falls into rivers.
- A volcano with a low and wide cone.
- A very large crater caused when the top is blown off a volcanic cone.

A volcano

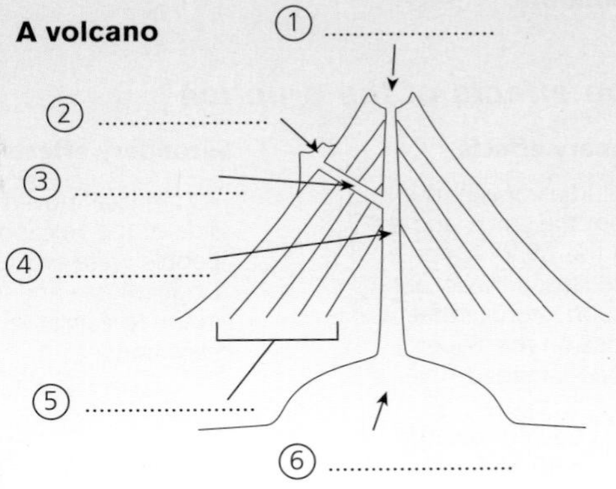

pyroclastic flow
main vent
lahar
magma chamber
shield volcano
side vent
caldera
crater
layers of ash and lava
dormant volcano
secondary cone

Case study: a tectonic hazard event in an MEDC

Mount Etna, Sicily

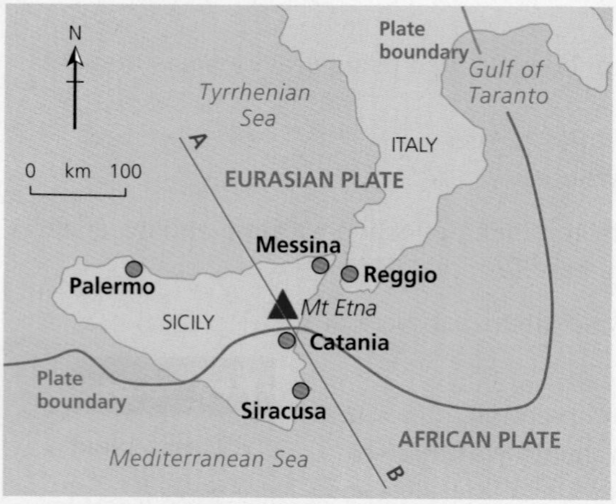

The location of Mount Etna volcano

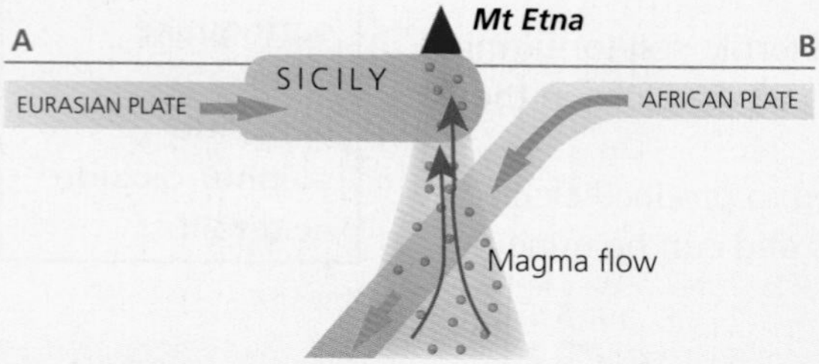

Simplified cross-section A to B across the plate boundary

Fact file

- Etna has been building for 35,000 years. The main cone is surrounded by a number of side vents.
- In October 2002, violent earthquakes heralded a major eruption. Lava flowed down the slopes, there was pyroclastic activity and volcanic ash was thrown as far as Tripoli in north Africa. The eruption ended in January 2003.
- Mt Etna is Europe's highest and most active volcano. It stands 3329 metres (10,922 feet) above sea level.
- More than 25% of Sicily's population live on its slopes.
- Sicily's main source of income is from tourism and the vineyards and orchards that cover the rich volcanic soils of its slopes.

Cause of eruptions

Mt Etna lies on a subduction zone where the African tectonic plate is sliding beneath the Eurasian plate. Here, magma with its attendant gases is rising to the surface to produce the volcano.

The effects of the eruption

Primary effects

Columns of ash rose from the crater and fell on the city of Catania and closed the local airport. Agricultural crops on the slopes were damaged

Lava flowed towards the south and to the north causing damage to tourist buildings, e.g. Rifugio Sapienza

Earthquakes measured 3–4 on the Richter scale. Santa Venerina and Guardia were damaged

Secondary effects

As the lava moved down the side of the volcano some people were evacuated from Linguaglossa and fires started in the forests at lower elevations

Tourist facilities on the sides of the volcano were closed; the tourist station at Piana Provenzana on the northeastern flank was destroyed and that of Rifugio Sapienza to the south was badly damaged

Schools and businesses were closed as ash covered a wide area

The earthquake damaged houses in Santa Venerina

Ash and smoke caused air traffic to be diverted, the airport at Cantania was closed and roads on the flanks of the volcano were also closed

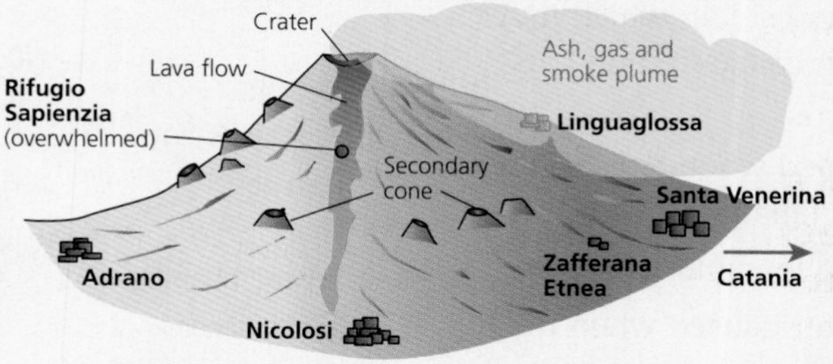

Reducing the impact of the eruption

- Mt Etna is an active volcano which erupts regularly and has done so for many centuries. Because of the fertility of the volcanic soils the land on its sides and around the volcano has been settled on and farmed for centuries.
- The volcano has been closely studied and monitored, eruptions are predicted and warnings issued. People understand the need to heed the warnings and act accordingly. Consequently there was no loss of life in the 2002–2003 eruption.
- During the 1991–1993 eruption the authorities built earth barriers and used explosives to successfully divert the lava flow away from the town of Zafferana. In 2002 they built barriers to divert lava from towns and a research station.
- Sicily is part of Italy, which is an MEDC. Communications are good, warning procedures are well rehearsed, emergency services are well trained and effective hospital facilities are available. As a result, the impact of any eruption is reduced.

Examination question

At 5 a.m. on 26 December 2003 disaster struck the ancient Iranian city of Bam. An earthquake measuring 6.8 on the **Richter scale** had its **epicentre** at Bam. The earthquake was caused by the movement of two **tectonic plates**, the Arabian Plate and the Eurasian Plate. The earthquake destroyed over 70% of the buildings in the city of Bam and the death toll reached 26,271. Thousands of people were left homeless.

(a) Read the extract carefully:

(i) Explain the meaning of the words in bold. *(3 marks)*

(ii) Describe *one* way in which the movement of plates can cause earthquakes. *(4 marks)*

(iii) Give two reasons why people live in hazardous locations. *(4 marks)*

(iv) Suggest three precautions which can be taken to reduce the effects of earthquakes. *(6 marks)*

Case study: *tectonic natural hazard*

Foundation tier:

(b) For an earthquake you have studied:

(i) Name the earthquake.

(ii) Describe the effects of the earthquake.

(iii) Why are the effects of earthquakes more damaging in less economically developed countries (LEDCs) than in more economically developed countries (MEDCs)? *(8 marks)*

Higher tier:

(b) Using examples from earthquakes you have studied, describe and explain why earthquakes in less economically developed countries (LEDCs) are often more destructive than those in more economically developed countries (MEDCs). *(8 marks)*

Exam tip

Notice the command words 'describe' and 'explain'. Make sure you know what they are asking you to do — see 'Command words' on pages 5 and 6. Make sure you write about a named example in question (b) to gain maximum marks.

Tropical storms: a climatic hazard

Tropical storms, known as **hurricanes** around the Caribbean Sea and the USA, are intense low-pressure systems. They are known as **typhoons** in east Asia and **cyclones** in south Asia.

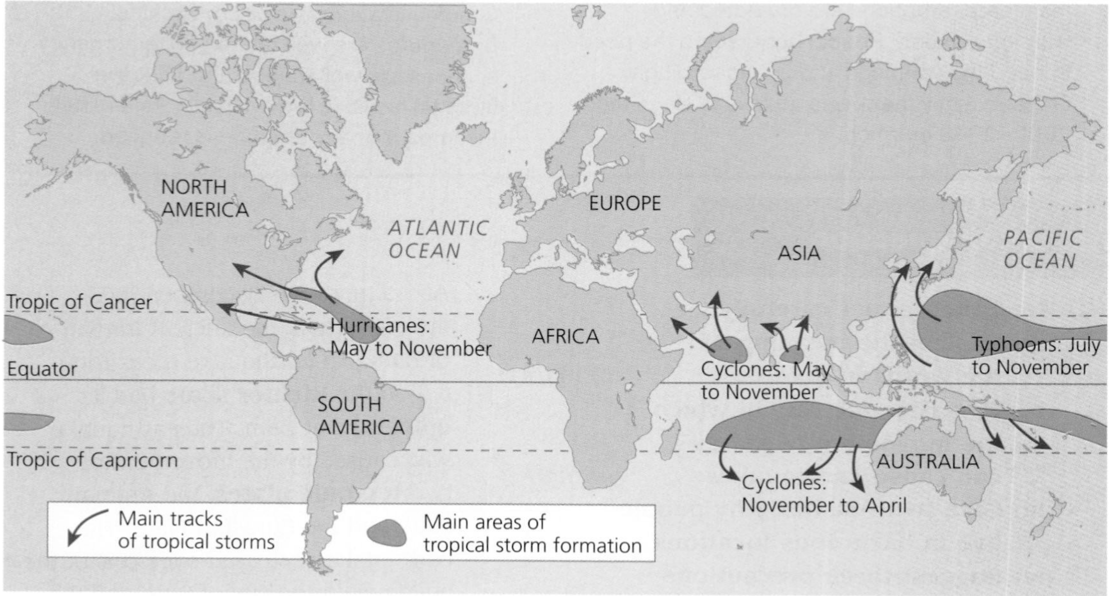

Where tropical storms are formed

Tropical storms develop in the tropics above a warm sea (warmer than 27°C). They form at the hottest time of the year, between May and November in the Northern Hemisphere and between November and April in the Southern Hemisphere. Typical tracks of tropical storms are shown on the map.

Formation of tropical storms

Tropical storms need warm tropical seas to generate their energy. Warm, damp air spirals upwards towards the top of the hurricane, which can be 20 kilometres high. As the air rises it cools and water vapour condenses, forming thick

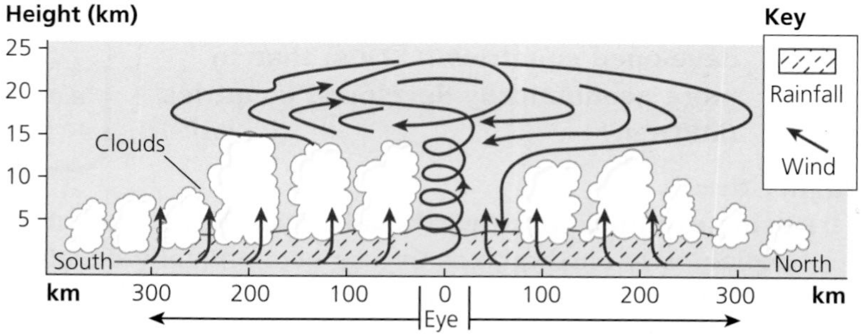

Cross section through a tropical storm

clouds, and producing torrential rain. The central area has very low pressure, but it is calm and clear and is known as the **eye of the storm**. Very strong winds blow outwards from the eye, reaching up to 300 kilometres per hour.

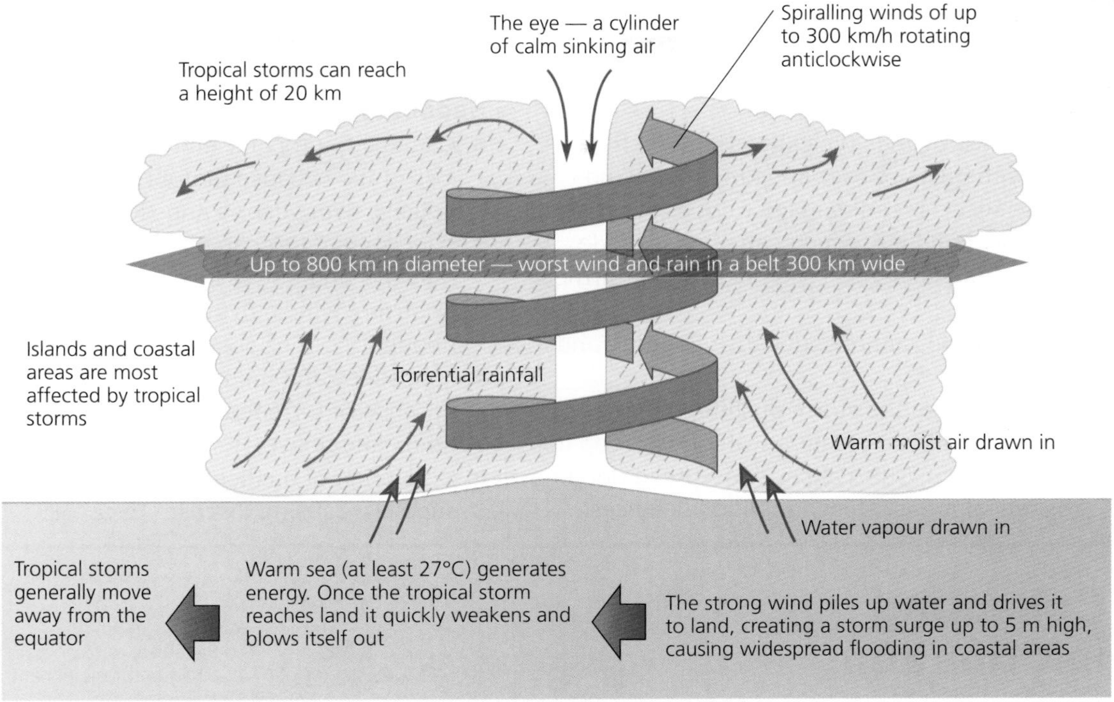
The eye — a cylinder of calm sinking air
Spiralling winds of up to 300 km/h rotating anticlockwise
Tropical storms can reach a height of 20 km
Up to 800 km in diameter — worst wind and rain in a belt 300 km wide
Islands and coastal areas are most affected by tropical storms
Torrential rainfall
Warm moist air drawn in
Water vapour drawn in
Tropical storms generally move away from the equator
Warm sea (at least 27°C) generates energy. Once the tropical storm reaches land it quickly weakens and blows itself out
The strong wind piles up water and drives it to land, creating a storm surge up to 5 m high, causing widespread flooding in coastal areas

Features of a tropical storm

Why people live in known tropical storm areas

People live in tropical storm areas in the full knowledge of the disruption to their lives such a storm may cause.

- It is where they were born and have lived all their lives.
- Good farm land, which is low-lying and in danger from storm surge, provides people with their livelihood, e.g. Bangladesh.
- Coastal ports for trade and fishing, e.g. New Orleans, provide people with work and with their livelihood.
- Attractive beach locations and tourist areas provide a pleasant environment in which to live, e.g. The Bahamas, Florida.

The impact of any hazard is related to the effect the hazard has on the lives of people. The impact is measured in terms of lives lost, buildings and industry destroyed, and the disruption a hazard causes.

People make what preparations they can in case a storm strikes. Such preparations are inevitably more thorough and effective in an MEDC which has the money and organisation to put plans into effect.

Key words

tropical storms
hurricane
typhoon
cyclone
eye of the storm

Impact of tropical storms

Primary effects

- Winds: with winds typically over 200 km/h and with gusts of over 300 km/h the wind causes widespread damage to trees, crops and insecure buildings which can even explode as pressure changes. Flying debris is a further hazard.
- Rain: intense rainfall can cause flooding and mudslides.
- **Storm surge:** the storm pushing the sea water towards land can create a wall of water up to 5 m in height. This causes widespread flooding to coastal areas, breaching flood defences.

Secondary effects

- Damage to property and consequent loss of life because of poorly built houses, flying debris and storm surge.
- Widespread dislocation of roads, railways and power lines caused by strong winds, flying debris and flooding.
- Landslides and mudslides may occur because of torrential rainfall, burying roads and houses.
- Disruption of medical and water supplies, causing disease.

Collecting information and prediction

Computers are used to collect, collate and monitor the information which is gathered to forecast the storm track and its intensity. In the USA this work is carried out at the National Hurricane Center. Based on this information, data from previous storms and computer simulations, the future track of a hurricane can be forecast quite accurately — but there is always the unexpected.

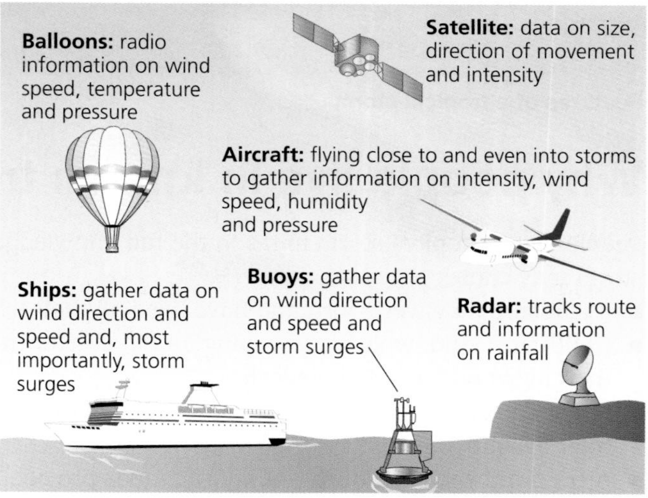

Collecting information and prediction

Measurement

The intensity of a tropical storm is measured on a scale of 1 to 5 known as the **Saffir-Simpson hurricane scale**. Storms become hurricanes when their average wind speed reaches 119 kilometres per hour (km/h).

For example, a category 3 hurricane would have a wind speed of 178–209 km/h and a storm surge height of 2.7 to 3.7 metres. A category 5 hurricane's figures would be (greater than) 249 km/h and (greater than) 5.5 metres, respectively.

Protecting people

How can people be protected from tropical storms?

Protecting people

Prediction: close monitoring by aircraft, satellites, ships and radar enables forecasters to monitor the route of the storm and predict its future course

Planning: authorities in areas prone to storms plan warning systems, evacuation routes and relief centres in safer areas, usually inland, e.g. USA

Education: publicity campaigns and teaching in schools can ensure that people know what to do in the event of a tropical storm, e.g. in Cuba

Defences: there are few defences against tropical storms but defence shelters have been built in areas susceptible to them, e.g. in Bangladesh

Building: building regulations can help to ensure that buildings are robust enough to withstand high winds; improve road systems to aid evacuation and strengthen flood defences

Key words

storm surge
Saffir-Simpson hurricane scale

Test yourself

1 **What is the main requirement for a tropical storm to develop?**

2 **What is the 'eye of the storm' in a tropical storm?**

3 **What is a storm surge and how is it formed?**

4 **Which of the following are primary effects and which are the secondary effects of a tropical storm?**
   (a) Landslides and mudslides
   (b) Very strong winds
   (c) Intense rainfall
   (d) Widespread dislocation of communications

5 **List four ways people can protect themselves from tropical storms.**

6 **List three ways in which data are collected for use in forecasting the track of a tropical storm.**

7 **Give three other names for tropical storms.**

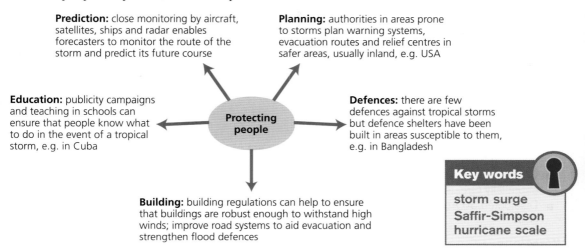

Case study: *a tropical storm — a climatic hazard event in an MEDC*

Hurricane Katrina, USA

Fact file

- Tropical storms like Katrina are called hurricanes in the Caribbean and the USA.
- Formed 23 August 2005 over the Bahamas.
- Highest wind speed 280 km/h (175 mph).
- It was a category 3 storm when it hit the Louisiana coast near New Orleans.
- Damage: most costly hurricane in history, estimated at US$200 billion.
- Lives lost: 1,836 people confirmed dead, 705 missing.
- 1,577 of the lives lost were in Louisiana, mostly in New Orleans.
- Greatest damage was in Louisiana, especially Greater New Orleans.
- Dissipated 30 August 2005 over central USA.

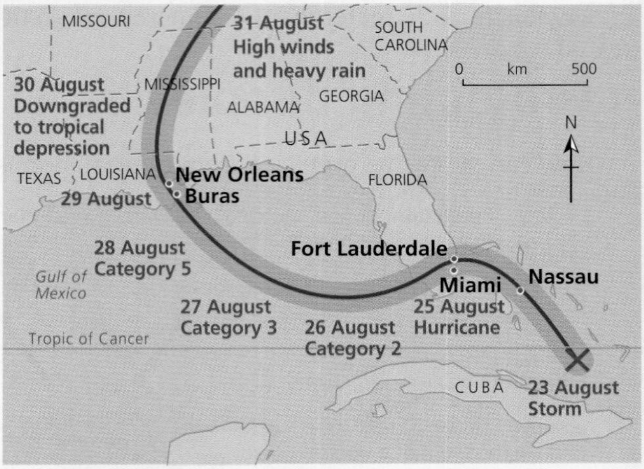

The path of Hurricane Katrina

Warnings

Warnings were given:

- National Hurricane Center tracked Hurricane Katrina and warned of its approach.
- The US Coastguard called up reserve members and made preparations.
- President Bush declared a state of emergency on 27 August.
- On 28 August the National Weather Service predicted devastating damage.
- On 28 August 1.2 million people were issued with evacuation orders. As a result motorways became jammed as people left the city, emergency centres were set up, including the Louisiana Superdome in the centre of the city, for those without transport and unable to leave.

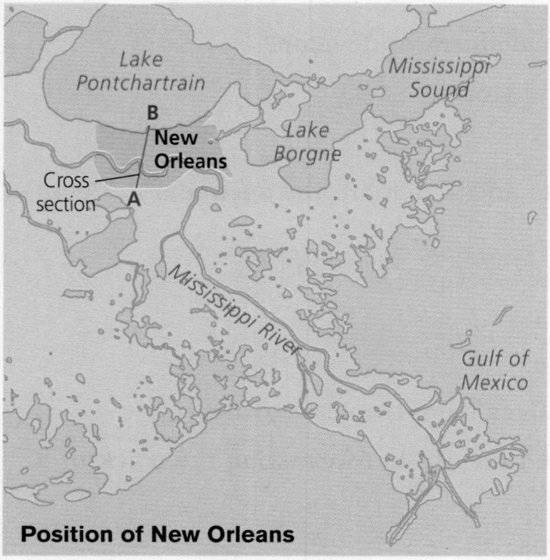

Position of New Orleans

Impact on New Orleans

Primary impacts

- High winds of 205 km/h (125 mph) up to 190 kilometres (120 miles) from the centre.
- Heavy rainfall of 200–250 millimetres, with a maximum of 380 millimetres in some parts of Louisiana.
- Storm surge in excess of 4.3 metres in some places.

Secondary impacts

- High winds caused trees to fall, blocking roads, and caused windows to be blown out and other damage to high-rise buildings.
- Heavy rainfall contributed to the rise in the water level of Lake Pontchartrain and made the flooding worse.

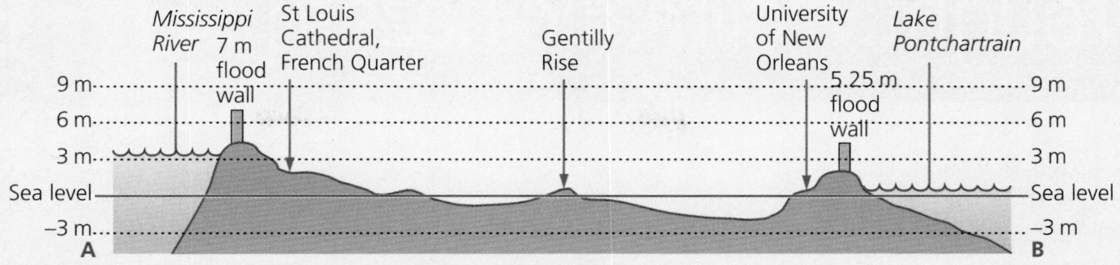

Cross-section of New Orleans A–B. The city lies between two floodwalls

- Storm surge caused the greatest impact. The levée system protecting low-lying New Orleans was breached in 53 places, causing widespread flooding across 80% of the city and loss of life. Bridges were washed away, electricity, gas and water services were destroyed or polluted. Houses were destroyed or made uninhabitable.

The response

The response was hampered by difficulties in communications caused by the widespread flooding. People were stranded without food and clean water, raising the possibility of the spread of disease.

Immediate

- Volunteers assisted those fleeing the storm with emergency housing.
- The coastguard rescued over 33,500 people.
- 58,000 National Guard personnel and additional police from other states were mobilised to help with law and order, rescue and clean up.

Follow-up

- The US Congress authorised US$62.3 billion in aid for the victims.
- Caravans (trailers) were provided for people to live in and hotel bills were paid for the homeless.
- Floodwaters containing sewage, toxic waste and oil were pumped out of the city into Lake Pontchartrain.
- Red Cross, Oxfam and other charities worked to support victims and re-establish communications.
- Over 70 countries pledged monetary aid and other assistance, including hygiene kits, food and mobile hospitals.
- Plans were made to strengthen the flood defences of New Orleans.

Factors affecting the impact of Hurricane Katrina

New Orleans lies below the level of both the Mississippi River and Lake Pontchartrain. The city should be protected by:

- **natural levées** — high earth banks built to withstand surges of water. There have been suggestions that these were not built to the highest standards
- the **low lying delta** which absorbs the storm surge and reduces its intensity and height. The effectiveness of the delta was reduced because:
  - the floodwalls and channels for shipping hasten the flow of the water to the sea and stop sediment being added to the delta
  - the delta wetlands have also been under attack from nutria (large burrowing rodents) which also eat the plant roots that bind the soil together, so eroding the delta
  - the extraction of oil offshore causes the sea floor to sink, taking the delta sediments with it

In New Orleans, although evacuation and other emergency plans were implemented, flooding hampered rescue efforts and caused loss of life. In other areas and towns affected by the hurricane, emergency plans worked effectively and many fewer lives were lost.

Drought: a climatic hazard

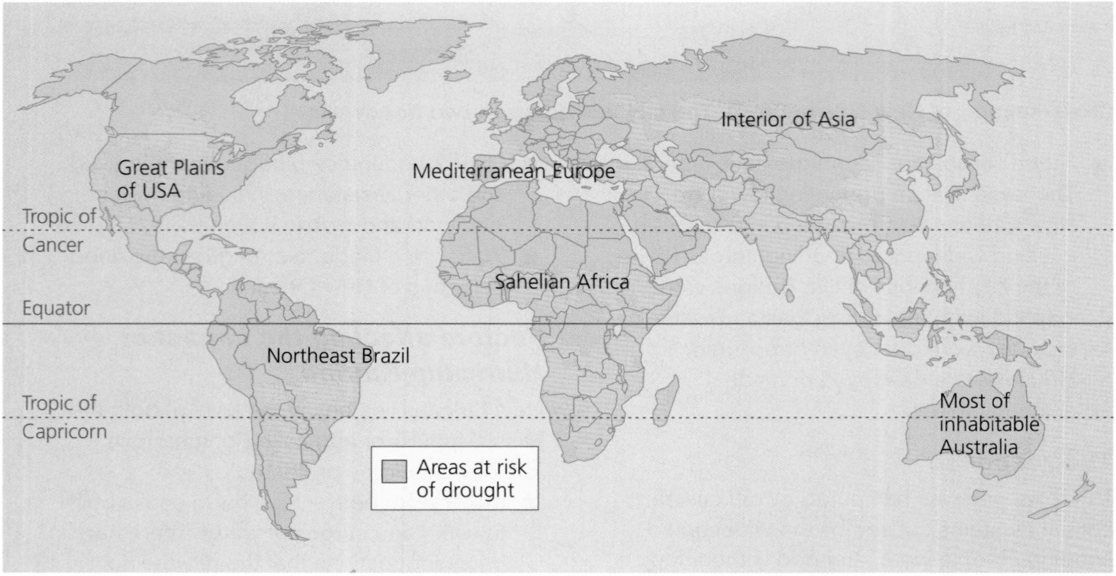

Global distribution of drought risk areas

What is a drought?

A drought is a prolonged period without enough **precipitation** to support people, animals or crops.

Droughts usually occur where rainfall is **seasonal** with alternating wet and dry seasons. If the expected rains fail then there will be no water for crops or for grazing.

The impact of droughts can be devastating. They develop slowly and the emergency may not be recognised until it is severe.

Prolonged droughts in desert areas are expected and have little impact.

What causes a drought?

Droughts are caused by a shift in the expected weather patterns on a global scale which causes seasonal rainfall to fail. For example:

- In Britain, high-pressure systems block the rain-bearing westerly storms from the Atlantic. If this happens for prolonged periods a drought occurs.
- In the Sahel region of Africa, unusually high temperatures in the Southern Hemisphere and Indian Ocean can prevent moist air at the equator from moving north. If this happens, the seasonal rains may fail, causing drought.

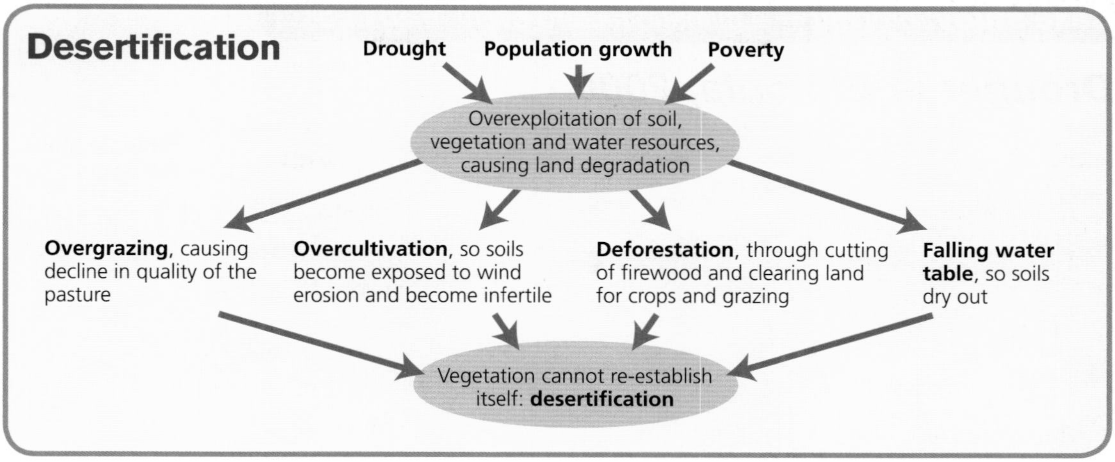
Desertification
Drought
Population growth
Poverty
Overexploitation of soil, vegetation and water resources, causing land degradation
Overgrazing, causing decline in quality of the pasture
Overcultivation, so soils become exposed to wind erosion and become infertile
Deforestation, through cutting of firewood and clearing land for crops and grazing
Falling water table, so soils dry out
Vegetation cannot re-establish itself: desertification

Droughts: the effects

Primary effects: insufficient water for people, animals and crops

Secondary effects:

- may not be immediate but as the period of drought lengthens, the severity deepens
- farmers may not be able to find grazing and water for their livestock, which become more susceptible to disease, grow thin, weaken and die
- greater pressure is placed on the available grazing and woodland, which can destroy them for future generations
- farmers' crops fail because of lack of water. This, and the effect on livestock, can cause widespread famine
- the water table drops and springs, streams and eventually rivers dry up
- drinking water becomes scarce, wells and reservoirs dry up and people become more susceptible to disease
- drought and heat provide the perfect conditions for wildfires, such as those which often devastate parts of California, Greece and Australia

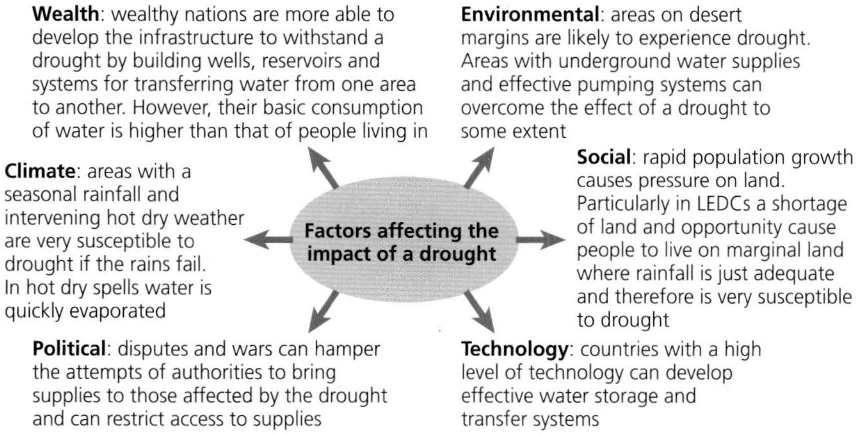
Wealth: wealthy nations are more able to develop the infrastructure to withstand a drought by building wells, reservoirs and systems for transferring water from one area to another. However, their basic consumption of water is higher than that of people living in
Environmental: areas on desert margins are likely to experience drought. Areas with underground water supplies and effective pumping systems can overcome the effect of a drought to some extent
Climate: areas with a seasonal rainfall and intervening hot dry weather are very susceptible to drought if the rains fail. In hot dry spells water is quickly evaporated
Factors affecting the impact of a drought
Social: rapid population growth causes pressure on land. Particularly in LEDCs a shortage of land and opportunity cause people to live on marginal land where rainfall is just adequate and therefore is very susceptible to drought
Political: disputes and wars can hamper the attempts of authorities to bring supplies to those affected by the drought and can restrict access to supplies
Technology: countries with a high level of technology can develop effective water storage and transfer systems

Key words

- precipitation
- seasonal rainfall
- overgrazing
- overcultivation
- deforestation
- desertification

Case study: *drought — a climatic hazard event in an LEDC*

Drought in Ethiopia, 2006

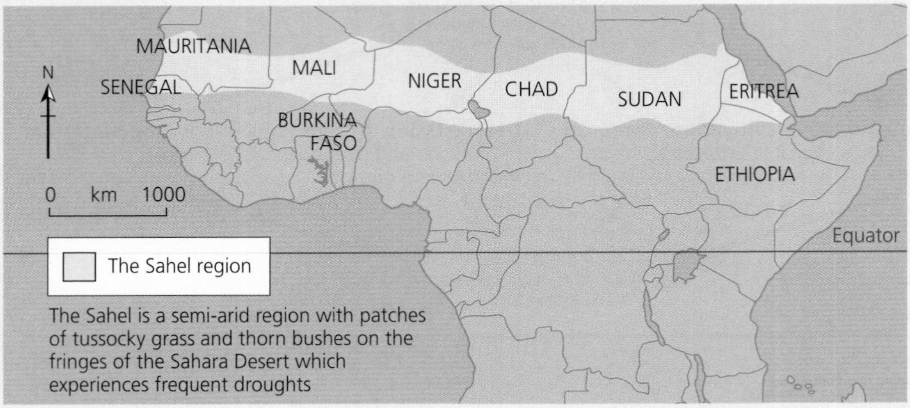

Ethiopia and the Sahel

Fact file

- Population: 77.7 million, of which over 85% live in rural areas. Life expectancy 43 years.
- Area: 1.2 million square kilometres — almost five times the size of the UK.
- Wealth: one of the five poorest countries in the world.
- Landscape: rugged highlands, particularly to the centre and west of the country, with lower lands to the east and south.
- Climate: seasonal rainfall with long rains from March to the end of May and short rains from September to October.
- Agriculture: traditional farming methods with little technology.
- Employment: over 80% of the population live in rural areas and rely on farming for a living.
- Civil war: between 1998 and 2000 a violent civil war between two regions stopped any development, cost millions of pounds and left much of the country devastated.
- Health: 2–3 million people suffer from HIV/AIDS. Only 20% of the population have access to safe drinking water.
- Aid: Ethiopia gets more relief aid than any other LEDC. Drought conditions have developed every few years since the 1970s. In 1984–1985, 8 million people died.

Causes of the drought

- Ethiopia is a poor country and does not have the infrastructure of dams and wells to cope with a drought.
- In the south and southeast of the country the recent rainy seasons were too little, too late and too erratic for water sources to be replenished, crops to grow and grazing to be re-established.
- The Earth is getting warmer and with this warming comes extremes of rainfall and temperature and more droughts, particularly in areas in the Sahel which are especially prone to drought.

Rainfall in Ethiopia

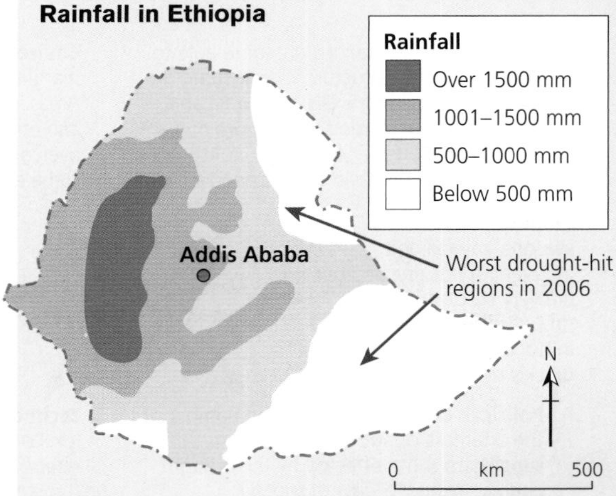

Impact of the drought

Primary effects

- The short and long rains failed causing a water shortage particularly in the south and east of the country.

Secondary effects

- The grazing for livestock died, the condition of the animals worsened and their value decreased. As the drought continued they became diseased and died.
- Water levels in ponds, deep wells and boreholes dropped and, as sources dried up, water supply became critical.
- Crops failed because of lack of water and the people were forced to eat their seed crop.
- People became weak through lack of food and, as their health deteriorated, diseases such as malaria, diarrhoea and intestinal diseases became common, particularly in children.
- Herdsmen moved their animals in search of water and grazing. Animals died along the way and there were reports of herders fighting over scarce grazing.
- Schools closed as children left to help their families survive or fell ill.
- The government turned to the outside world for help and aid agencies estimated that more than 6 million people needed assistance.

The response

Immediate

- Food aid was distributed at feeding shelters where fresh drinking water supplies were also provided.
- At emergency centres aid agencies provided vaccines to fight disease and oral rehydration salts to help fight the effects of diarrhoea.
- Aid agencies supplied community water tanks to help store water supplies, which needed to be brought in by tanker in some areas.

Medium term

Once the initial emergency was over people needed to return to some form of normality, so there was a need to:

- provide seeds of dry-region plants such as maize and beans, and tools such as hoes and sickles, so that the farmers could plant crops to feed the people
- provide new breeding stock for the livestock herders to build their herds
- repair and maintain existing water points and provide experts to assess needs and advise on supply
- return schools to normality to provide security and continuity for children

Long term

Long-term aid was required to combat the effects of future droughts, and to develop more sustainable support for the people to enable them to survive future droughts, such as:

- the development of low-cost systems to conserve and make best use of water supplies, for example:
  - building low walls across slopes to slow water runoff and provide greatest benefit to the soil
  - building micro-dams to store water
  - ensuring the maintenance and repair of existing systems and wells
  - planting trees to help conserve moisture in the soil
- the development and planting of drought-resistant strains of crops such as maize, beans and wheat to combat drought conditions
- working with the people and their government to develop a strategy that can be put in place if further droughts occur. For example, the pastoralists for whom their livestock is their wealth could be encouraged to reduce their herds if drought strikes, in order to feed people, conserve feed and maintain a core herd for breeding once the drought has passed

Drought hazard in an MEDC

Impact

Droughts usually have less impact in an MEDC than in an LEDC because:

- MEDCs have the resources to build water transfer systems whereby water can be moved from one part of the country to another in time of need. For example, in Britain large reservoirs such as Carsington Water in Derbyshire store water during wet weather and guarantee water supply during dry weather. In the USA water is moved from the Colorado River many miles to California to irrigate crops.
- People's livelihoods and food supply do not depend on specific rainfall as they do in an LEDC.
- People can be warned at the onset of drought and the consumption of water can be reduced for non-essential uses, e.g. hosepipe bans for watering the garden and for watering golf courses.
- In the long term an MEDC has the capacity to put measures in place, including building and pipeline projects, to provide a sustainable response and lessen the impact of a drought in the future.

Key word

bushfires

Drought hazard and bushfires

Australia is the driest continent in the world and much of its centre is desert. To the southeast and southwest the country has a Mediterranean-type climate with hot dry summers and cooler wetter winters. This area is prone to drought if the winter rainfall fails to come or is too low.

In the southeast, the government has developed an irrigation system from the great Murray and Darling rivers which flow through the region. Boreholes are widespread, bringing water from aquifers which is also used to water livestock.

In 2010 a severe drought that had been continuing for 4 years was reaching crisis point. It was believed to be the worst drought in 100 years. In this MEDC the government responded in the following ways:

- It banned the use of water supplies for swimming pools and watering lawns and parks.
- It put in place a scheme for Goulburn, south of Sydney, to become the first city in Australia to use recycled waste water and treated sewage water for drinking.
- It decided to build more desalination plants to turn sea water into drinking water for cities like Sydney, Perth and Melbourne.
- It helps struggling farmers by providing cash as income support and information on how to adapt to the drought. It also helps farmers who wish to sell their farms to retrain.

The effect of the hot, dry conditions is to dry out the vegetation cover, leaves and material on the ground. This can easily be ignited by lightning, the glass of a dropped bottle magnifying the Sun's rays, a dropped cigarette or deliberate action by people, causing devastating **bushfires**.

The Australian bushfires in 2009

Fact file
- Fires swept through the State of Victoria in Australia in February 2009
- 179 people were killed
- 2,000 homes were destroyed
- Whole townships were wiped out, including Marysville and Kingslake
- Experts believe that 1 million native animals died

Causes
- The 'Big Dry' — 12 years of drought
- Extremely hot weather in January and February (summer in Australia)
- Strong winds
- Vegetation and leaf litter were tinder dry
- Fires started by people and lightning

Impact on people
- In addition to deaths up to 7,000 people lost their homes
- Smoke haze grew over Melbourne. Those with asthma, the old and very young were warned to stay indoors

Impact on the environment
- Large numbers of animals died including 3,000 koalas
- Large areas of eucalyptus forest destroyed

Prevention
- Not allowing undergrowth and leaf litter to build up by controlled burning
- Creating long gaps between the trees, known as escape clearways, to stop fires
- Improving the early warning system
- Changing the policy that encourages people to stay and defend their property
- Developing escape routes, fireproof bunkers and better firefighting methods
- Reviewing the building code that allows people to build in the bush

Impact on towns
- Tourist industry devastated — hotels and holiday homes burnt down
- One-fifth of the population of Marysville killed
- Marysville burnt down

The future
- Average Australian temperatures are expected to rise by 0.6°C to 1.5°C in the next 20 years because of global warming
- It is expected that the number of days of very high and extreme fire danger will rise by 25% by 2020

Test yourself

1 Complete the following using the correct word or words from those in brackets.
A drought is a (*short/prolonged*) period without enough precipitation to support people. Droughts usually occur where rainfall is (*year round/seasonal/monthly*). The impact of droughts can be devastating as they develop (*without warning/slowly/quickly*). During a drought the water table will (*rise/drop*) and the wells will (*dry up/be easier to use*). If farmers cannot find enough grazing, their livestock will (*become more susceptible to disease/grow thicker hides*). Bushfires can be caused by (*light winds/tinder dry vegetation/cool weather*); they can be prevented by (*building long gaps between the trees/allowing leaf litter to build up*).

(a) List four main factors affecting the impact of a drought.
(b) List three causes of **desertification**.
(c) Give two reasons why droughts will have less impact in MEDCs than LEDCs.
(d) Give four ways in which the effect of droughts can be reduced.

Examination question

Cyclone Nargis crossed the Bay of Bengal and hit the coast of Myanmar. The majority of the 130,000 deaths were caused by the storm surge. With waves reaching 7.6 m high it flooded long distances inland along the densely populated southern coastline of Myanmar.

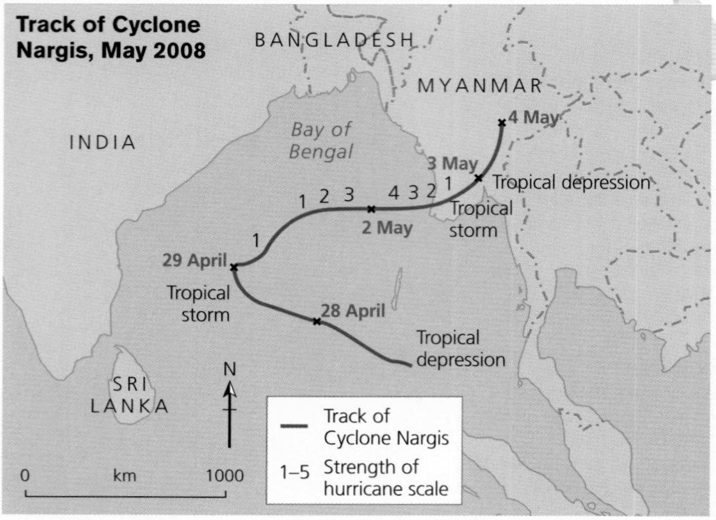

(a) Study the map and text describing Tropical Cyclone Nargis.

(i) Describe how the strength of the tropical storm shown on the map changes between 29 April and 4 May. *(2 marks)*

(ii) Explain why each of the changes you describe occurred. *(4 marks)*

(b) (i) Give three major features of a tropical storm. *(1 mark)*

(ii) Explain how a storm surge occurs. *(2 marks)*

(c) Explain two ways in which the impact of tropical storms is more severe in LEDCs than in MEDCs. *(4 marks)*

(d) Describe three ways in which people can be protected from the full effects of a tropical storm. *(4 marks)*

Exam tip

Notice that even the 'short' questions can carry 4 marks. Make sure you include at least four main points in your answer to gain these marks. Always answer questions using complete sentences.

Case study: *climatic hazards*

Foundation tier:

(e) For a climatic hazard you have studied

(i) Name the climatic hazard.

(ii) Describe the short- and long-term impacts of the event.

(iii) Explain what measures are being taken to reduce the impact of the hazard in the future. *(8 marks)*

Higher tier:

(e) (i) Name a type of climatic hazard and the location where it occurred.

(ii) Explain the short- and long-term impacts of the event and how human activities affected the impact of the natural hazard. *(8 marks)*

Theme 4

Economic development

Measuring development

As well as wealth, measuring development must take into account many other factors. Development depends also on factors such as education and healthcare. We would not say that a country was developed unless people living there were able to enjoy a reasonable quality of life.

Country	HDI rank	GDP (PPP US$)	Birth rate (per 1000)	Life expectancy (years)	Urban population (%)	Adult literacy (%)	People per doctor
Nigeria	158	1,969	43	47	47	72	3,571
Egypt	123	5,349	21	72	43	66	1,851
Peru	78	7,838	21	71	76	87	1,370
Argentina	49	13,238	19	75	91	98	370
South Korea	26	24,801	10	79	82	97	1,076
UK	21	35,130	13	79	90	99	623
USA	13	45,592	14	79	82	99	408
Japan	10	33,632	9	82	79	99	613

The table shows six **indicators of development** which can be used to decide how developed a country is. Other indicators, such as infant mortality rates or access to safe water, can also be used.

Study the table carefully and try to think about what these indicators tell you about a country. For example, why are **literacy** rates so low in some countries and how might this affect the country's development?

- Why are birth rates in **MEDCs** generally much lower than in **LEDCs**?
- How could **life expectancy** in LEDCs be increased?

Measures of development

Gross Domestic Product (**GDP**) is a measure of the wealth of a country and can be used as an indicator of the relative wealth between different countries.

Advantage: clear measure of economic development

Disadvantage: does not indicate social well-being or **quality of life**

Human Development Index (**HDI**) was developed by the UN to describe both economic and social well-being within countries. The HDI is based on three factors: life expectancy, literacy and GDP.

Advantage: the HDI is thought to give a better indication of the quality of life enjoyed by the people in a country

Disadvantage: it includes only three indicators: are they the best indicators? Should there be more indicators used to give a more accurate picture?

GDP and HDI can give very different results. For example, in 2006 South Africa was ranked 70th in the world for GDP but 125th on the HDI table, and Cuba ranked 97th on the GDP table yet was ranked 48th on the HDI table.

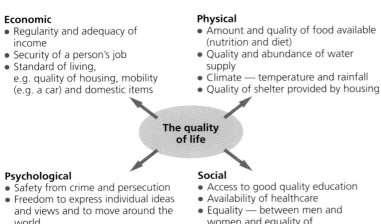

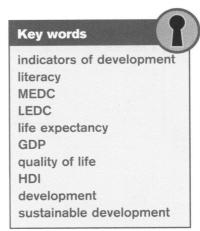

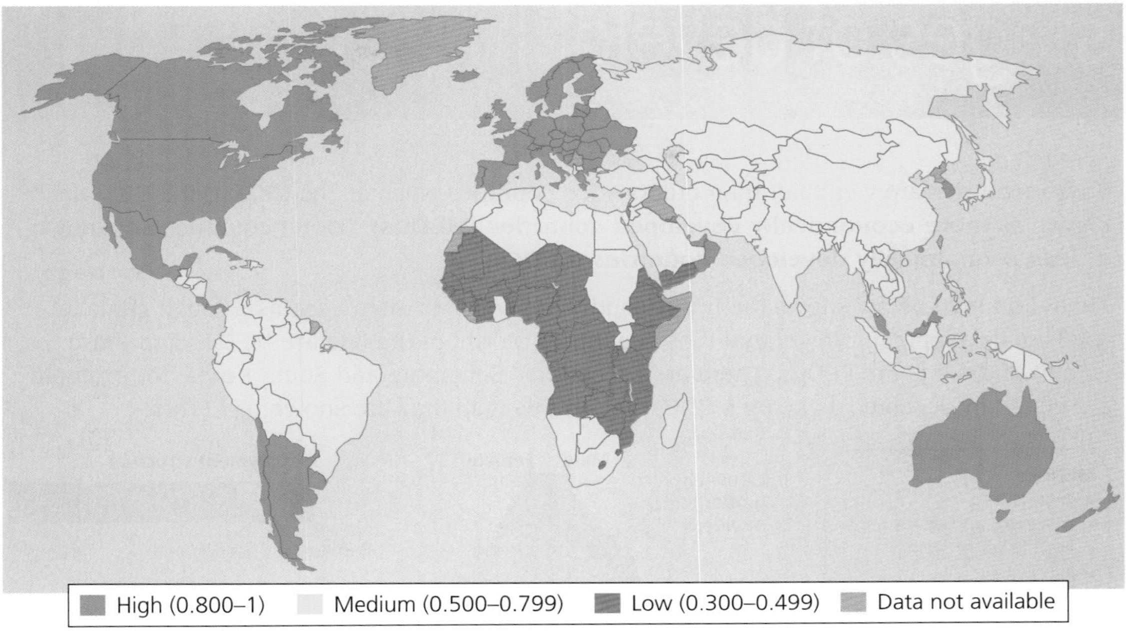

High, medium and low HDI levels of economic and social wellbeing

Development occurs where there is improvement in the factors (such as income, health, security and diet) that increase a person's quality of life.

Sustainable development is where development that benefits people has no long-term costs. For example, building of renewable energy power stations, such as solar power, which do not use scarce resources or contribute to pollution and global warming.

The quality of life

Economic
- Regularity and adequacy of income
- Security of a person's job
- Standard of living, e.g. quality of housing, mobility (e.g. a car) and domestic items

Physical
- Amount and quality of food available (nutrition and diet)
- Quality and abundance of water supply
- Climate — temperature and rainfall
- Quality of shelter provided by housing

Psychological
- Safety from crime and persecution
- Freedom to express individual ideas and views and to move around the world
- Access to a fair justice system and human rights
- Content with circumstances in which one lives (happiness)

Social
- Access to good quality education
- Availability of healthcare
- Equality — between men and women and equality of opportunity e.g. in access to employment
- Support of family and friends

Key words

indicators of development
literacy
MEDC
LEDC
life expectancy
GDP
quality of life
HDI
development
sustainable development

Why countries are at different stages of development

Some countries are wealthier than others. Rich countries, such as the UK and the USA, are known as **more economically developed countries (MEDCs)**. Poorer countries are known as **less economically developed countries (LEDCs)**.

The world map below shows the north–south line, which is often used as a rough guide to levels of development. In general those countries 'north' of the line are MEDCs and those 'south' of the line are LEDCs. There are exceptions. Singapore and South Korea, for example, are generally accepted as being MEDCs, yet on this map they are shown as LEDCs.

MEDCs
- High GDP
- Low birth and death rates
- High levels of literacy
- Export mainly manufactured goods
- Most people have access to safe water and sanitation

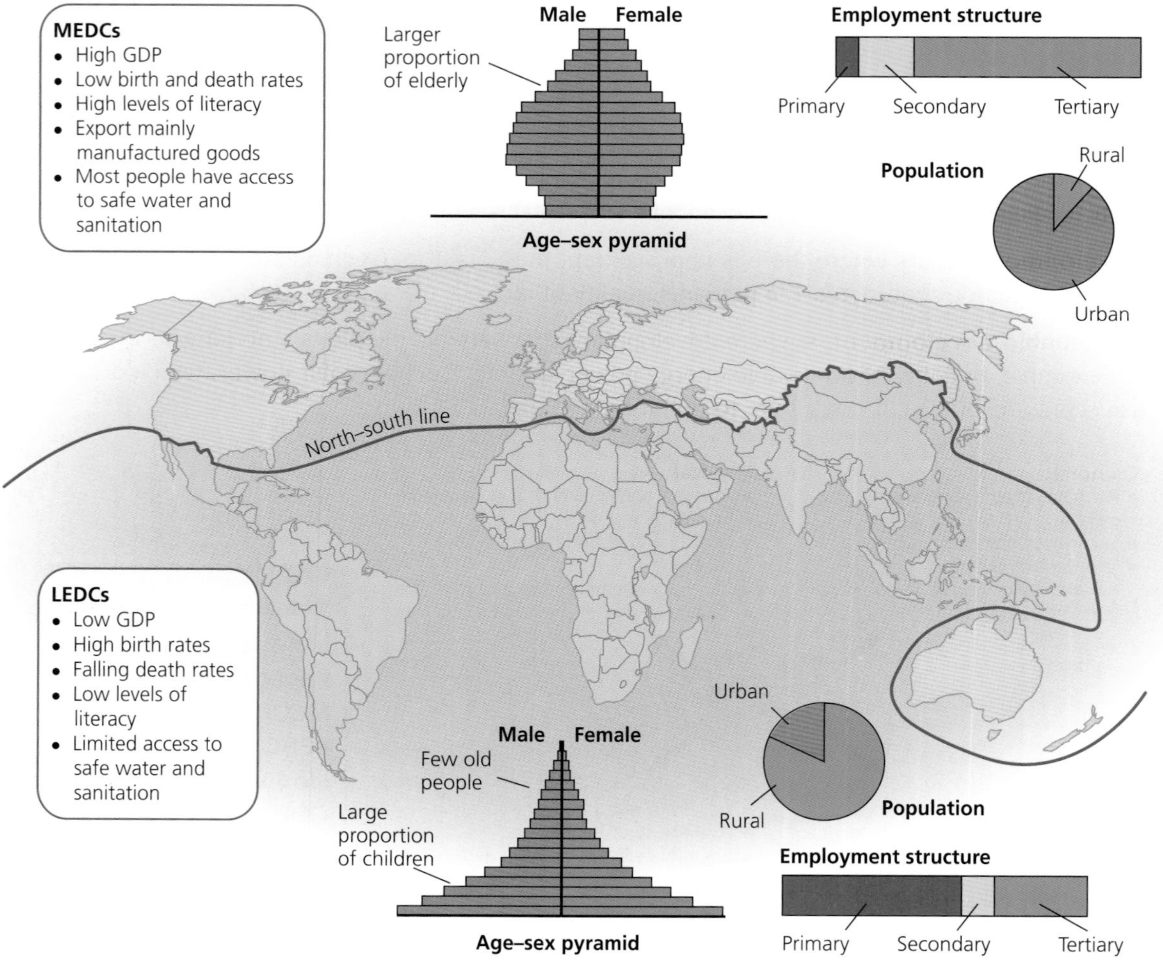

LEDCs
- Low GDP
- High birth rates
- Falling death rates
- Low levels of literacy
- Limited access to safe water and sanitation

Why are countries at different stages of development?

There are a variety of reasons why countries are at different stages of development. For any one country there are usually several different reasons.

Why countries are at different stages of development

Human resources
A highly skilled and well-educated workforce in a country with few natural resources can become wealthy by using imported resources to produce high-value goods to sell to others

Population
In many LEDCs population growth has been rapid. With high birth rates the economy has not been able to grow fast enough to provide the housing, education and food supplies the people need

Many LEDCs have an imbalance in the structure of their population, with large proportions of children who are dependent on the older members of society to look after them

Natural resources
Natural resources such as oil, timber, metallic ores and good soils are all valuable to people. Those countries that possess them are able to increase their wealth by using them to manufacture goods to sell to other countries or selling them directly to others

History and politics
Many LEDCs were former colonies controlled by MEDC European countries such as Britain, France and Germany. The colonial powers used them as a source of raw materials and a market for their manufactured goods, so stifling development in the colony

Poor government in many LEDCs has hampered development. Corrupt governments and tribal conflict have resulted in civil war and little development

Geography
The location of a country can influence its development. Access to the sea is important for trade and landlocked countries are dependent on others to ship their goods abroad. Small island countries lack the resources to develop

Uncertain climate can hamper progress if the country is constantly battling with drought and flood

Newly industrialised countries (NICs)

1950	1	2	3	4	5	2000
Agriculture 24% Industry 30% Services 46%	Few natural resources Prosperous farming Processing rice and sugar cane	Developing and expanding textiles and clothing manufacture Cheap, low-tech production process using craft skills people already have	New industries based on cheap labour and raw materials Plastics as by-product of oil Electrical goods from components made in Japan and USA	New heavy industries: shipbuilding, steel-making, oil-refining, chemicals	Better-educated workforce develops new skills Quality improved, prices kept low Electrical goods, e.g. televisions New infrastructure including motorways	Agriculture 4% Industry 42% Services 54%

The growth of industry in Taiwan

Examples of **NICs** are the southeast Asian countries of South Korea, Taiwan, Singapore and Hong Kong. These so-called 'tiger economies' have broken out of the cycle of low profits and lack of investment. They have succeeded in this through:

- investment from other countries, mainly Japan and USA, because of the motivated workforce working for low wages
- investment in infrastructure (road, rail and ports) for communications
- early industries such as clothing, shoes and paper products which relied on relatively cheap raw materials, abundance of cheap labour, low technology and the use of craft skills which the workforce already possessed
- over time the workforce developed new skills to make electrical goods, computers and cars

Key words
- MEDC
- LEDC
- NIC

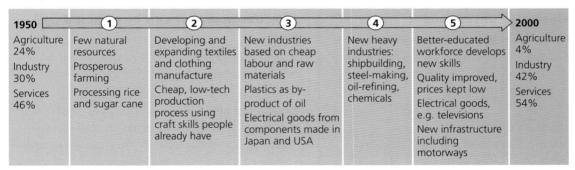

Aid

International aid is the transfer of resources such as money, equipment, food, training or skilled people from MEDCs to LEDCs. Aid is given to help LEDCs either in an emergency or for long-term economic development.

Different types of aid are given in different ways.

Bilateral aid Aid given directly from one government to another in the form of money, training, technology, food or other supplies. In some cases this aid is **tied aid**, which means it has conditions attached which will usually benefit the donor country.

Multilateral aid Aid which comes from a number of different governments or organisations. It is usually arranged by an international organisation such as the World Bank or the United Nations (UN). These organisations usually give to large-scale projects.

Non-governmental aid Organisations such as Oxfam and Save the Children run projects all over the world, many of which are small-scale. They also help to organise **emergency aid** after disasters. These **non-governmental organisations** (**NGOs**) raise their money through donations and from government grants.

Sustainable development

Sustainable development considers the needs of future generations as well as the needs of people today. The Earth and its resources should be handed over to the next generation in the same condition as they are in today.

Sustainable development:

- should be able to be sustained in the long term
- will involve local people who have knowledge of the technologies and access to the resources they need to maintain the development
- will use resources which will not run out

Key words

multilateral aid | emergency aid
non-governmental aid | bilateral aid
sustainable development | tied aid
non-governmental organisation (NGO)

How development can be affected by aid

Many LEDCs have benefited from international aid, but their development may also be a result of other factors.

Benefits of aid

- Aid used to provide new technology and machinery can result in more jobs, and enable farmers to grow more crops and increase exports.
- New infrastructure projects providing new roads, power sources, clean water and sanitation improve the health and well-being of the people.
- Small-scale projects improve people's quality of life, self-esteem and maintain local culture.

Shortcomings of aid

- Large capital-intensive projects may have unforeseen environmental and social consequences.
- The aid, particularly in the case of large projects, may not benefit the poorest people.
- Often much foreign aid is bilateral aid, which is tied to joint projects and trade agreements.

Case study: *a large-scale aid project in an LEDC*

The Narmada Project in India

Advantages	*Disadvantages*
✔ large quantities of electricity generated	✘ local people oppose the scheme
✔ large areas of farmland irrigated	✘ over 1 million people will have to leave their homes as their villages are submerged
✔ flooding by river controlled	✘ the land offered to displaced people is poor
✔ fishing in the lakes created by the dams	✘ environmental damage to 24,000 hectares of forest — the replanting cannot replace lost habitats
✔ water for domestic and industrial use	✘ benefits will be slight, e.g. 123,000 hectares of land irrigated but 90,000 submerged
✔ 36,000 hectares of new forest planted to replace what is lost	✘ dams may silt up quickly, reducing storage capacity of reservoirs
	✘ the region is an earthquake zone

Large-scale aid projects such as this are criticised because they:

- cause long-term environmental damage
- permanently destroy local ways of life
- increase inequality
- do not benefit the local community

The Narmada Project has been very controversial, particularly the building of the Sardar Sarovar Dam. This dam was originally to be 80 m high but in a series of rulings the government gradually raised its height until in March 2006 it ruled that the height be raised to almost 122 m, flooding vast areas of additional land.

The 'Save Narmada Movement' has protested strongly on behalf of those displaced to ensure their rights are safeguarded. The supreme court ruled that they should be satisfactorily rehabilitated before construction began.

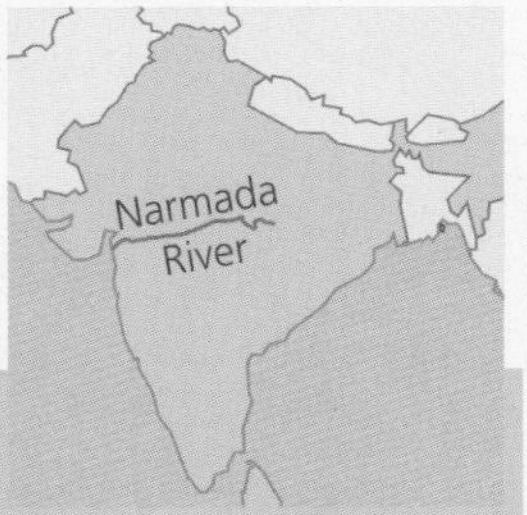

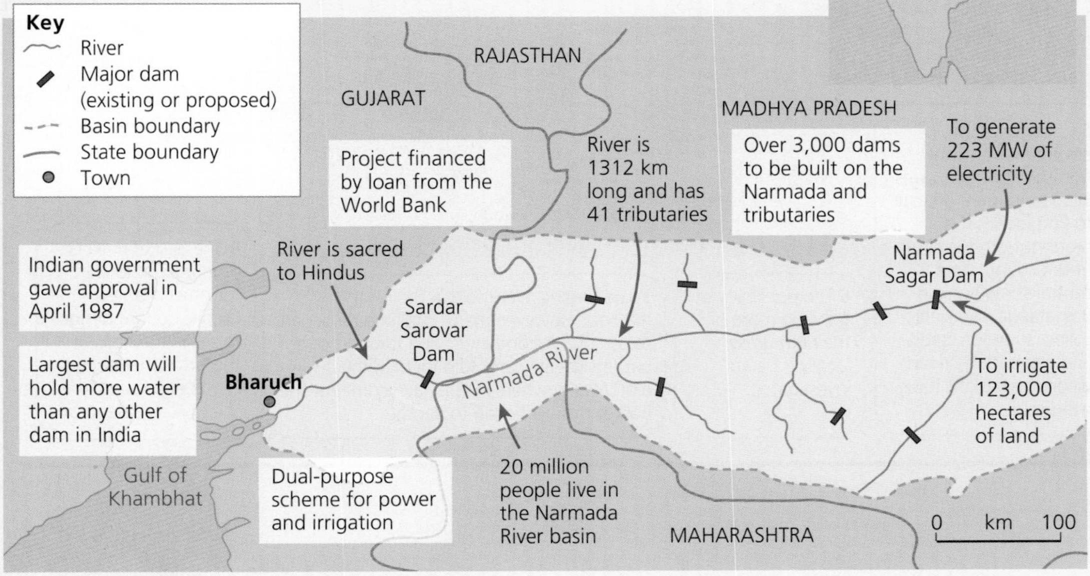

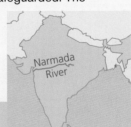

Case study: *a small-scale aid project in an LEDC*

Basket weaving in Ghana

Sustainable aid brings economic benefits without causing environmental, social or psychological damage for future generations. Many aid organisations believe that small-scale, locally based projects are more sustainable than large-scale projects such as the Narmada Project.

Small-scale projects improve people's quality of life, self-esteem and maintain local culture.

1 The north of Ghana is mainly dry savanna land. The people are mainly subsistence farmers. The average income is below internationally recognised poverty levels with earnings of less than US$1 per day

2 The people had a long tradition of basket making using the native savanna grasses

3 With an aid grant from Oxfam villagers were offered training in business management and encouraged to develop their basket-making skills and set up small businesses

4 Within 5 years the basket weavers' network had 408 members — 330 women and 78 men

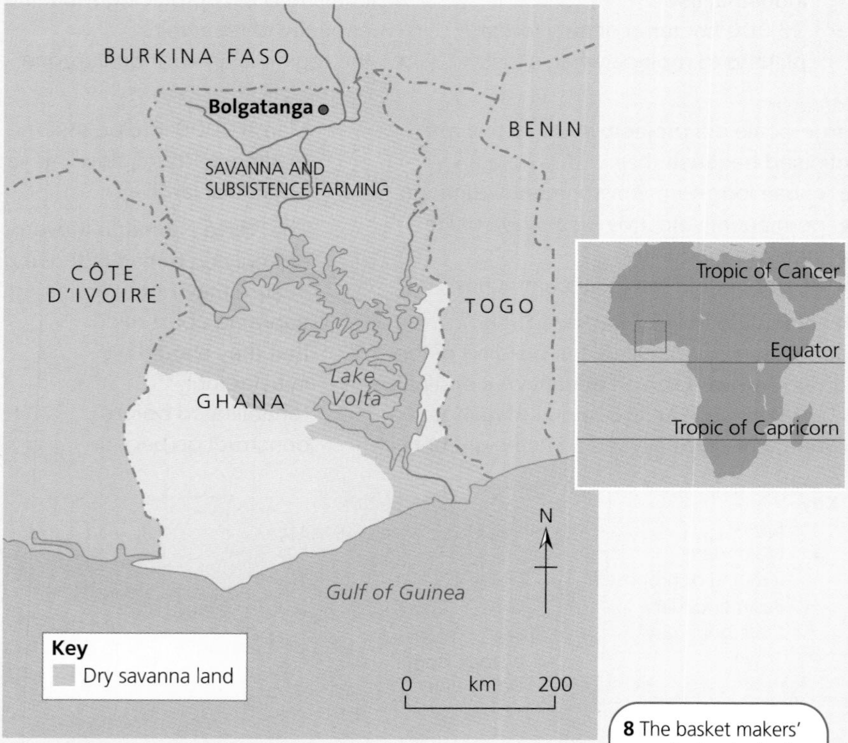

5 Changes in the environment meant that the right grass was becoming more difficult to find locally. It is brought from the south of the country to a central straw bank in Bolgatanga. Weavers collect straw on credit and pay back the loan when the baskets have been sold

6 The possibility of growing more of the right grass locally is being investigated

7 The baskets were sold in the local markets. However, many local people began to use plastic containers, and traders bought baskets cheaply and sold them in the markets elsewhere in Ghana. Income for the basket makers began to decline

8 The basket makers' organisation looked for other markets. Oxfam bought designs to sell in its shops in the UK. Other outlets have been found in the USA and Denmark

Test yourself

1 **What is the main advantage of measuring development using HDI rather than GDP?**

2 **Give two differences between MEDCs and LEDCs.**

3 **What are NICs? Give two reasons for their rapid growth.**

4 **Match these terms to the correct definition:**

bilateral aid	short-term immediate relief after a disaster
tied aid	aid given with conditions, usually to benefit the donor
NGO aid	aid given by one government to another
emergency aid	development projects run by charities such as Oxfam
multilateral aid	aid arranged by international organisations, e.g. the World Bank

Examination question

Indicators of development

Country	GDP (PPP US$)	Birth rate (per 1,000)	Life expectancy (years)	Urban population (%)	Adult literacy (%)	People per doctor
Argentina	13,238	19	75	91	98	370
Japan	33,632	9	82	79	99	613
Peru	7,838	21	71	76	87	1370

Study the table of indicators of development.

(a) Which of the three countries is a 'more economically developed country' (MEDC)? *(1 mark)*

(b) List what you consider to be the four main features which indicate a person's quality of life, giving reasons for your choice. *(4 marks)*

Which three indicators of development would you consider to give the best indication of 'the quality of life' in a country? Give reasons for your answer. *(3 marks)*

(c) Give three features of 'newly industrialised countries' (NICs). *(5 marks)*

(d) Explain what is meant by 'sustainable development'. *(3 marks)*

Case study: *an aid project in an LEDC*

Foundation tier:

(i) **Name and locate a development project in an LEDC you have studied.**

(ii) **Outline how the project was paid for, including the type of aid given. Describe the impacts the project has had on people and on the environment.** *(8 marks)*

Higher tier:

(e) Using a named example of a development project in an LEDC which you have studied, describe how the project has been funded and evaluate the success of the project. *(8 marks)*

Employment structure

All jobs can be divided into four main employment sectors:

Primary sector The gathering of the raw materials from which things are made. Raw materials occur naturally and can be:

- naturally occurring in the Earth, such as rocks which are quarried, oil which is drilled, or coal which is mined
- grown, such as farm crops or wood from forestry
- collected, for example fish from the sea, or rubber as latex from trees

Secondary sector Uses raw materials to make products, for example steel from iron ore. This may then be the raw material for another factory making cars.

Tertiary sector Does not make anything, but provides a service, such as retailing (shops), transport, teaching, hospitals, the police and civil servants.

Quaternary sector Very new research and development industries involved in biotechnology, information technology or communication.

Formal and informal sectors

Formal sector

(95% of workers in the UK)

- regular waged employment (long hours and relatively low pay)
- in factories or offices
- working for an employer
- many employers are foreign multinational companies

Informal sector

(Up to 40% of the economy in an LEDC, e.g. flower selling, shoe shining, bicycle repair, delivery services, small craft industries)

- irregular, insecure income
- in small workshops or on the streets
- self-employed and finding work wherever the opportunity arises
- no job security
- no money to invest in improvements, so job cannot develop

How employment structure varies between countries

The balance between the employment sectors described above is shown by the percentage of people employed in them. This is known as the **employment structure**.

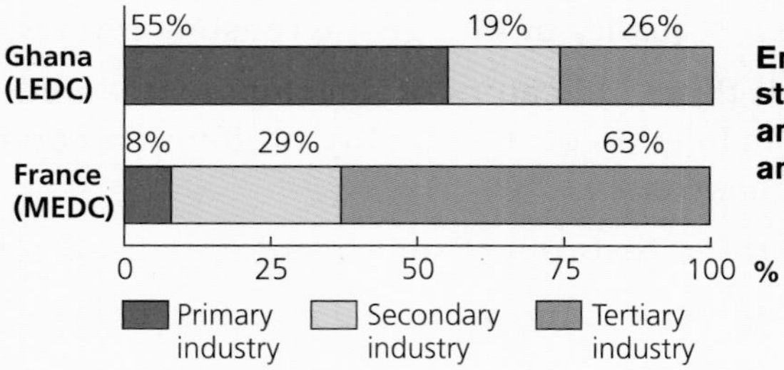

Employment structure in an LEDC and an MEDC

Employment structure is often used to show the level of development in a country. LEDCs have a large percentage of people in primary industry (generally farming). In MEDCs the largest percentage is employed in tertiary industry. As countries become more developed the percentage of primary industry falls and that of secondary and tertiary industry increases.

In NICs the secondary and tertiary sectors are rising. As farming becomes mechanised this sector becomes smaller and there may even be a small quaternary sector.

Comparing employment structures of different countries

A triangular graph is often used to compare the employment structures of different countries.

Each side of the triangle shows the data for one sector of employment, primary, secondary and tertiary, and has a scale of 0 to 100%. By following the direction of the arrows for Bangladesh you can see how it can be read.

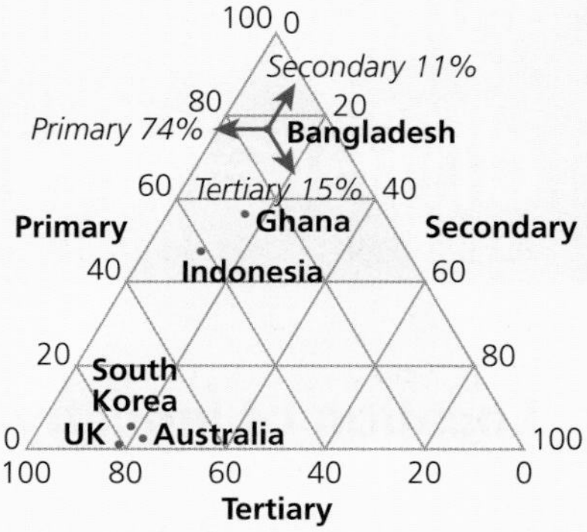

Triangular graph to compare employment structure

How employment sectors change over time

The graphs show how the employment structure is changing in the UK. Sixty years ago the primary and secondary sectors of industry were large. Today the tertiary sector dominates.

As an economy develops so the employment structure changes.

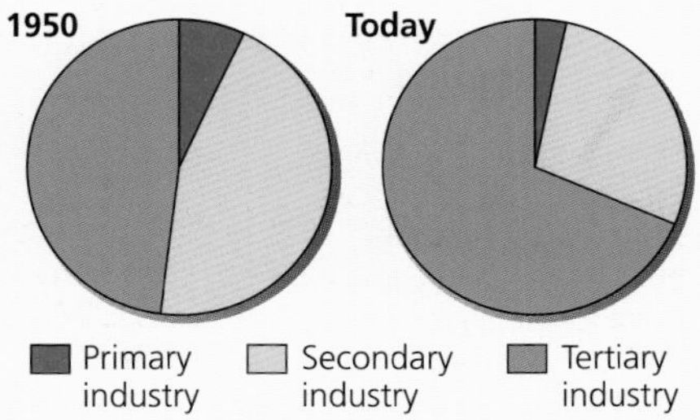

Change in employment structure in the UK

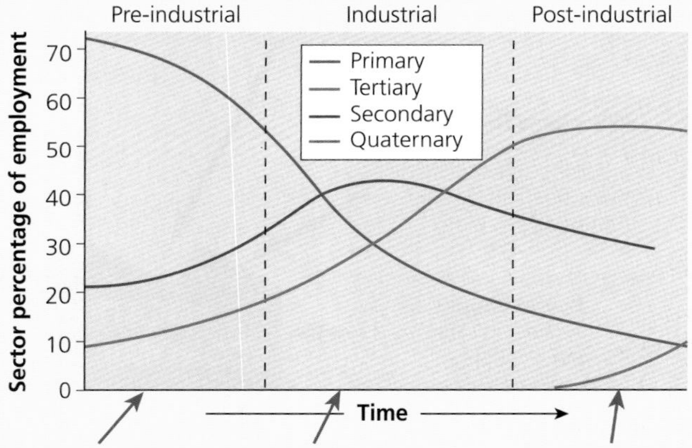

Most people work in agriculture, use animals or themselves as power, and sell their surplus in a local market, e.g. Bangladesh

People move to urban areas to work in the secondary sector in factories that use coal and later gas and oil as power. Goods are sold in local and overseas markets, e.g. South Korea

Tertiary employment dominates as wealth increases and education, health-care and leisure services increase, e.g. Australia

Sector model of employment change

How employment structures may change in the future

Some trends are apparent:

- To save money companies in MEDCs have **outsourced** their work to countries where costs and labour are lower.
- As LEDCs continue to develop, their primary sector will decline and their secondary and tertiary sectors will continue to grow.
- As NICs develop, their primary sector will decline in importance, their secondary sector will decline and their tertiary sector will grow and a quaternary sector develop.
- With new methods of manufacturing, the secondary sector will continue to decline in MEDCs.

Key words

- primary sector
- secondary sector
- tertiary sector
- quaternary sector
- formal sector
- informal sector
- employment structure
- outsourcing

What determines the location of different economic activities?

Location of industry

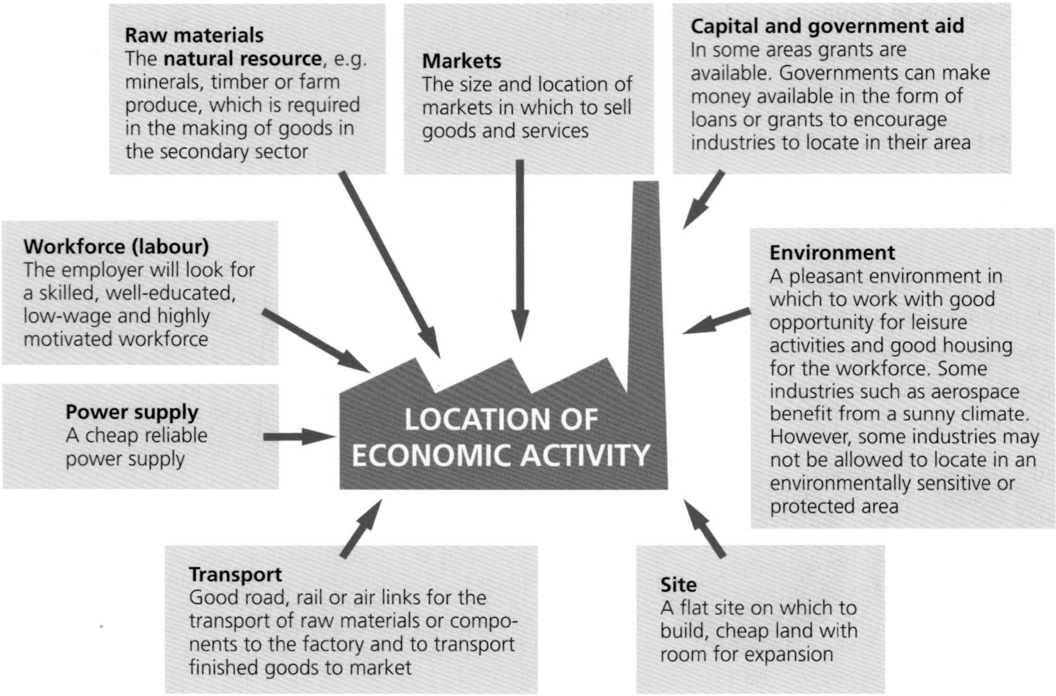
Raw materials
The **natural resource**, e.g. minerals, timber or farm produce, which is required in the making of goods in the secondary sector
Markets
The size and location of markets in which to sell goods and services
Capital and government aid
In some areas grants are available. Governments can make money available in the form of loans or grants to encourage industries to locate in their area
Workforce (labour)
The employer will look for a skilled, well-educated, low-wage and highly motivated workforce
Environment
A pleasant environment in which to work with good opportunity for leisure activities and good housing for the workforce. Some industries such as aerospace benefit from a sunny climate. However, some industries may not be allowed to locate in an environmentally sensitive or protected area
Power supply
A cheap reliable power supply
LOCATION OF ECONOMIC ACTIVITY
Transport
Good road, rail or air links for the transport of raw materials or components to the factory and to transport finished goods to market
Site
A flat site on which to build, cheap land with room for expansion

Not all these factors will be important for every industry looking for a suitable location. For example, the decision to site an aluminium plant in Ghana depended on an abundance of cheap power to smelt the raw material bauxite (the Akosombo Dam), the existence of raw materials (bauxite which is bulky and expensive to transport could be mined nearby) and a port to transport the aluminium (the plant was built on flat land at Tema on the coast). Other factors were less important.

Economic costs such as the transport of raw materials is an important factor. For example, the bulky raw materials for the iron and steel industry in the UK are mostly imported from abroad by sea, so the industry is mostly located at the coast.

Agglomeration: some industries cluster together to support each other. For example, highly technical industries and banking to whom the latest information is important or when component makers are supplying an assembly plant.

Primary industry

Primary industries, of necessity, are located where the raw materials are found. For example, farming where the climatic, soil and landscape conditions are right, fishing on the coast close to fish stocks and mining where the rock or minerals are found. The issue then is to bring all the other factors as economically and (more recently) environmentally acceptably as possible to the site and establish transport links to move the product to the processing plant.

It is often the case that the nearer the processing plant is to the source of raw material the more economical it is. So, for instance, in the case of quarrying for roadstone, bulky lumps of stone are crushed on site into gravel-sized pieces for easier transportation. Similarly, ore is often part-processed at the mine to lessen the bulk to be transported.

In the case of oil it is easier to transport the crude oil which is extracted from the ground for long distances to the refineries often halfway round the world for processing.

Case study: *location of a primary industry*

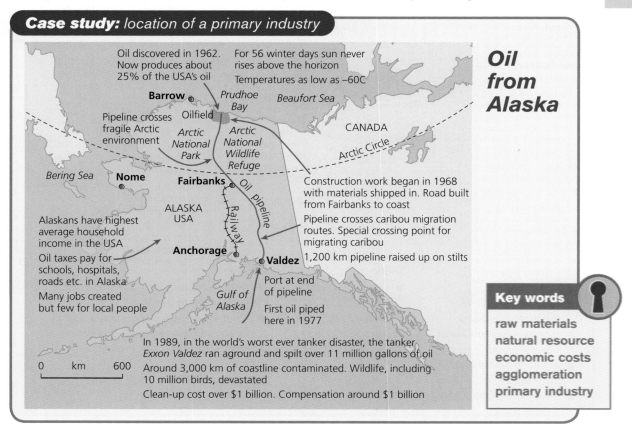

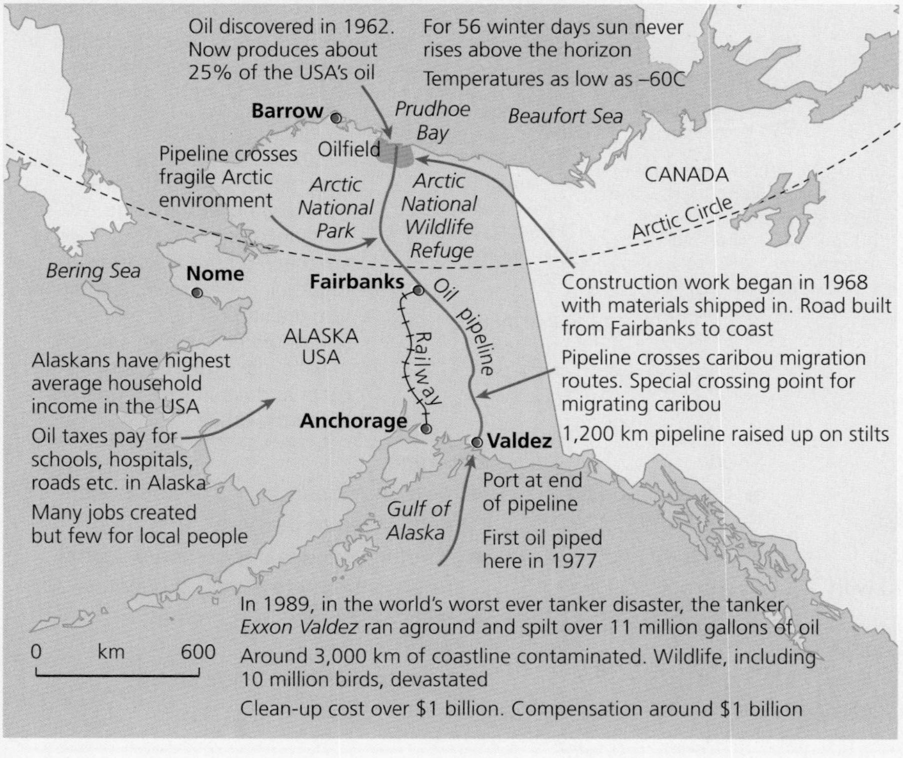

Oil from Alaska

Key words

- raw materials
- natural resource
- economic costs
- agglomeration
- primary industry

Secondary industry

Secondary industries range from traditional industries, such as the making of iron and steel from iron ore, coal and limestone or making clothes from wool and cotton, to modern industries, such as manufacturing cars and electrical goods.

Modern secondary industries are not tied to their source of raw materials or power and can locate where they like. Local and regional authorities may offer money and other incentives for a firm to locate in their area because they want the jobs and affluence the industry will bring.

Case study: *location of a modern secondary industry*

Toyota cars at Burnaston

Changes in south Derbyshire following the building of the Toyota factory

When industries which are linked are grouped together in one area it is called **agglomeration**

A **greenfield site** is one where no industry has been developed before

The opening of a new factory causes a boost to industry in the local area, e.g. more money spent in shops and new houses built, creating even more jobs. This is known as the **multiplier effect**

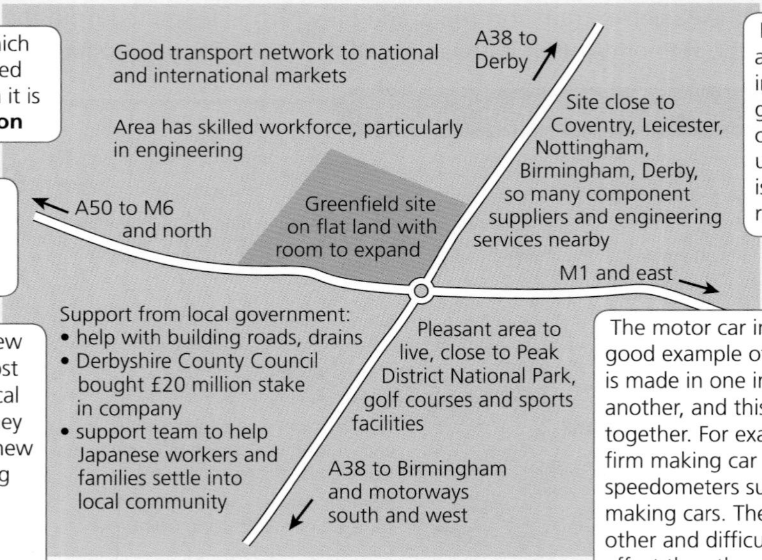

Good transport network to national and international markets
A38 to Derby
Area has skilled workforce, particularly in engineering
Site close to Coventry, Leicester, Nottingham, Birmingham, Derby, so many component suppliers and engineering services nearby
A50 to M6 and north
Greenfield site on flat land with room to expand
M1 and east
Support from local government:
• help with building roads, drains
• Derbyshire County Council bought £20 million stake in company
• support team to help Japanese workers and families settle into local community
Pleasant area to live, close to Peak District National Park, golf courses and sports facilities
A38 to Birmingham and motorways south and west

Industrial estates are areas in which industries are grouped, usually on the outskirts of urban areas. There is no residential or retail building

The motor car industry is a good example of **linkage**. What is made in one industry is used by another, and this links the industries together. For example, a component firm making car headlights or speedometers supplies the firm making cars. They depend on each other and difficulties in one will affect the other

Environmental

- farmland lost
- number of vehicles in the area increased
- factory developed with low buildings
- pollution and waste controlled
- trees screen the factory

Social

- Toyota sponsors local education, cultural and leisure activities

Economic

- new engineering and car component firms move to the area
- electricity and water companies gain new contracts to supply the factory
- over 2,500 new jobs created in the area
- new houses built in the area — building jobs

Traditional industry in the nineteenth and for the first half of the twentieth centuries often depended on the supply of raw materials and coal for power. These were bulky and expensive to move, so the industries located near to where they were found or could easily be imported.

Case study: *location of a traditional secondary industry*

Sugar factory at Cantley

Fact file

- The Cantley factory is owned by British Sugar plc which sells its sugar under the brand name Silver Spoon.
- The factory was established in 1912 and is still producing sugar on the same site.
- The sugar is made from sugar beet grown by 925 growers farming in Norfolk near to the factory.
- The beet is harvested and processed from September to December.
- At the peak time the number of employees increases from 112 to 155 and the factory works 24 hours a day every day of the week.
- 330 lorry loads of beet arrive at the factory each day in the harvesting season.

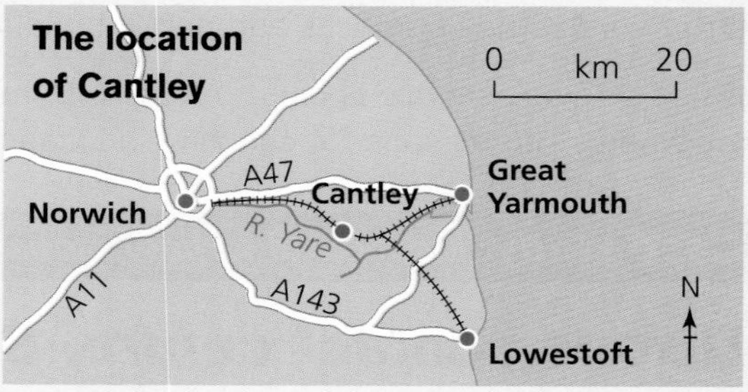
The location of Cantley
0 km 20
Norwich
A47
Cantley
Great Yarmouth
R. Yare
A11
A143
Lowestoft
N

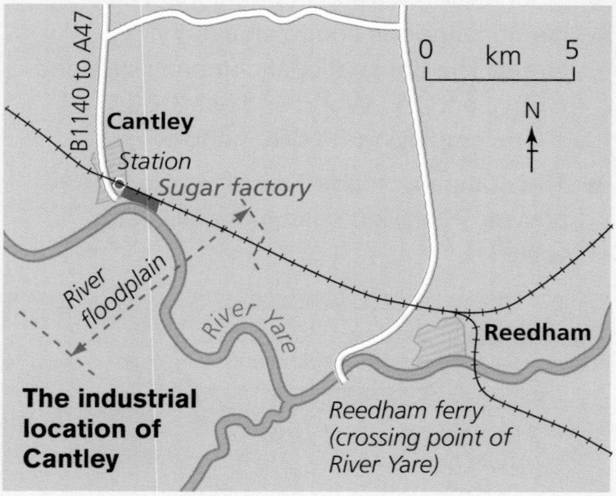
B1140 to A47
0 km 5
N
Cantley
Station
Sugar factory
River floodplain
River Yare
Reedham
Reedham ferry (crossing point of River Yare)
The industrial location of Cantley

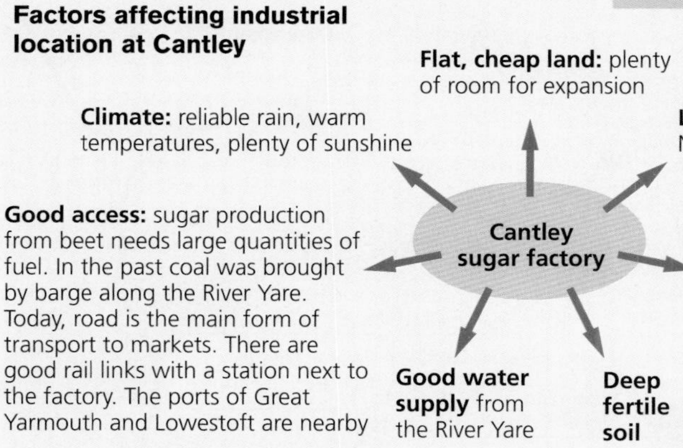
Factors affecting industrial location at Cantley
Cantley sugar factory
Flat, cheap land: plenty of room for expansion
Climate: reliable rain, warm temperatures, plenty of sunshine
Labour supply: the city of Norwich is 30 minutes away by car
Good access: sugar production from beet needs large quantities of fuel. In the past coal was brought by barge along the River Yare. Today, road is the main form of transport to markets. There are good rail links with a station next to the factory. The ports of Great Yarmouth and Lowestoft are nearby
Near to raw materials: the factory needs to be close to the growers as sugar beet is bulky and costly to transport
Good water supply from the River Yare
Deep fertile soil

Key words

secondary industry	industrial estate
agglomeration	linkage
greenfield site	traditional industry
multiplier effect	

Tertiary industry

Shops are one type of service or **tertiary industry**. There are also many people working in tertiary industries in offices and public buildings such as hospitals and schools.

A fast-growing service is call centres. With modern communications these can locate anywhere. The reasons for choosing a location are given in the case study below.

Case study: *location of tertiary industry*

Dixons contact centre, Sheffield

- Dixons Store Group International has electrical goods shops on every high street. The transnational company includes Currys and PC World and has 1,000 shops employing 35,000 people.
- The company needed a call centre to answer 70 million customer calls each year.
- The centre employs up to 1,200 people answering calls about customer concerns, products, deliveries and complaints.
- The centre opened in February 2000.
- The investment was £45 million.

Key words

tertiary industry
brownfield site

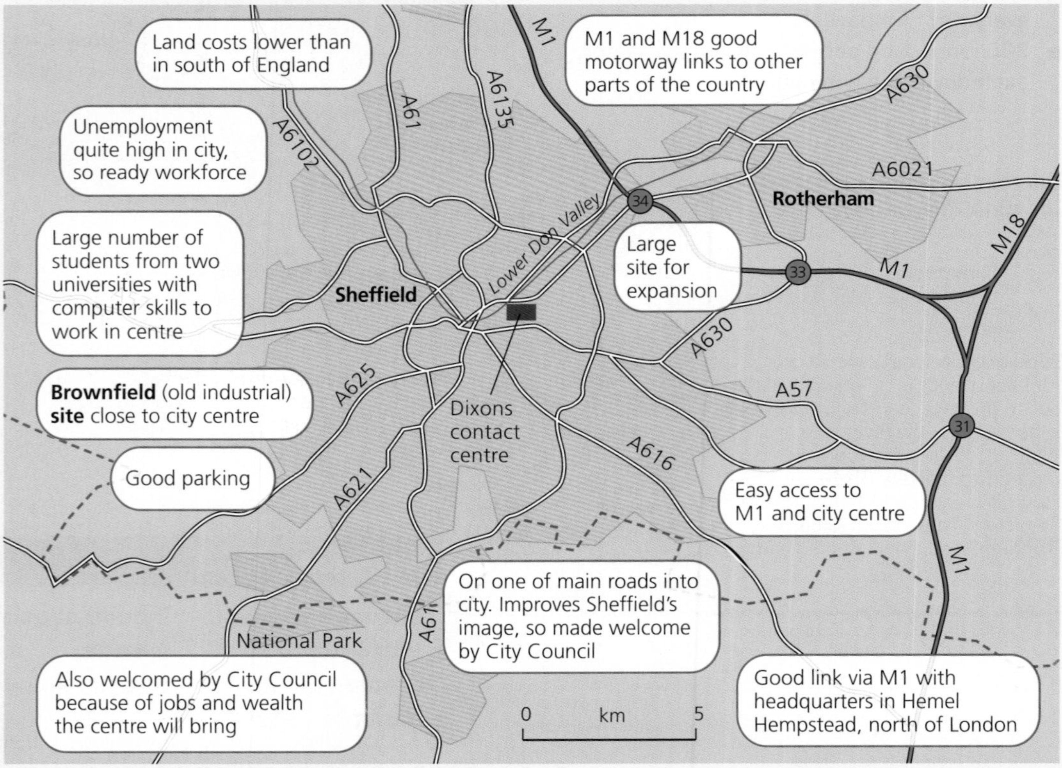

Land costs lower than in south of England
Unemployment quite high in city, so ready workforce
Large number of students from two universities with computer skills to work in centre
Brownfield (old industrial) site close to city centre
Good parking
Also welcomed by City Council because of jobs and wealth the centre will bring
M1 and M18 good motorway links to other parts of the country
Large site for expansion
Easy access to M1 and city centre
On one of main roads into city. Improves Sheffield's image, so made welcome by City Council
Good link via M1 with headquarters in Hemel Hempstead, north of London
Sheffield
Rotherham
Lower Don Valley
Dixons contact centre
National Park
M1
M18
A61
A6135
A6102
A630
A6021
A625
A621
A57
A616
34
33
31
0 km 5

Quaternary industry

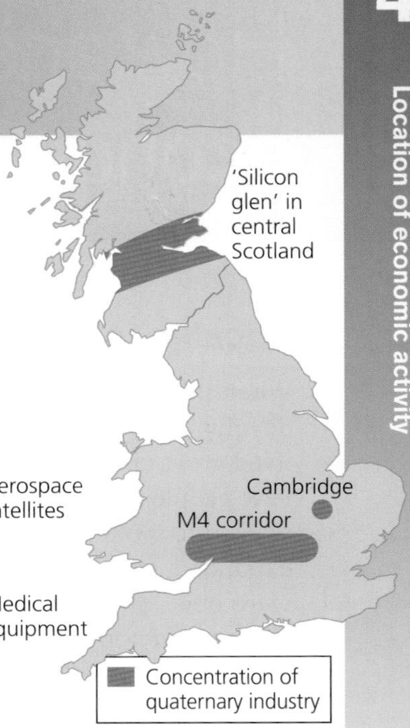

Quaternary or hi-tech (high-technology) industries are those linked to science and technology, e.g. making computers and mobile phones.

Quaternary industries are called **footloose** industries. They can locate anywhere because they do not depend on large quantities of raw materials or power.

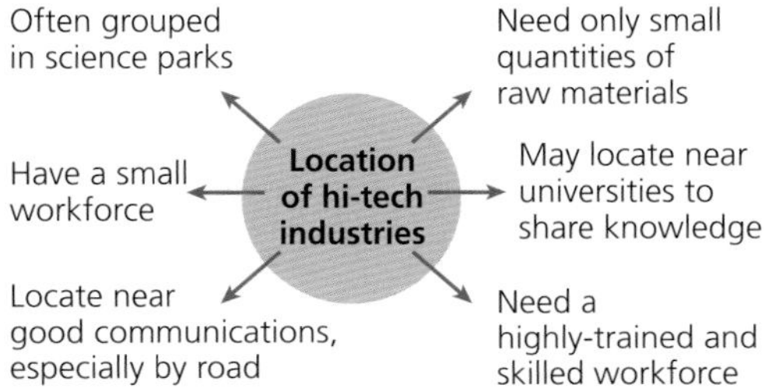

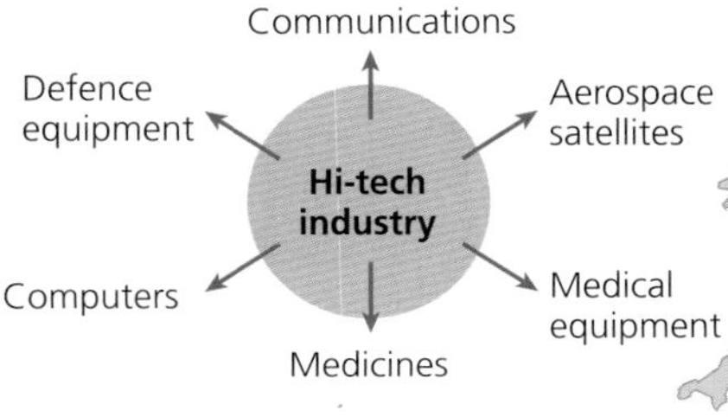

Case study: location of quaternary industry

Cambridge Science Park

- Started in 1970 on a disused site owned by Trinity College north of Cambridge.
- Activity restricted to research and development of new ideas employing scientists and skilled technicians.
- Buildings at low level, often unusual designs, with attractive landscaped areas between.
- Seventy organisations based at the park.
- Over half of the employees have a degree.

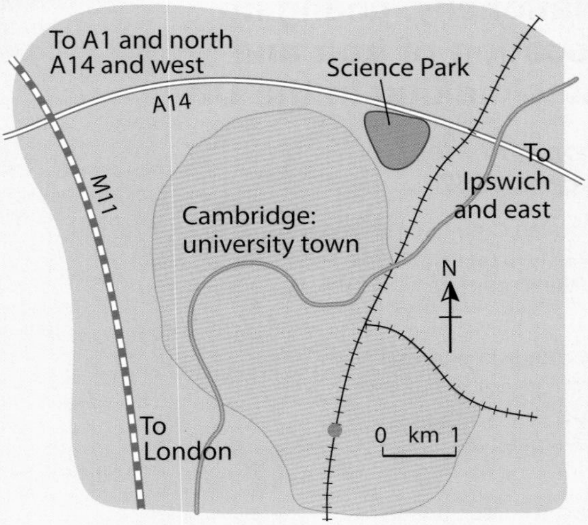

Reasons for locating in Cambridge

- Prestige: worldwide reputation of university for excellence in science and technology.
- Links with the university departments and the latest ideas.
- Highly skilled and qualified workforce.
- Attractive, well-landscaped site to create a good image and impress clients.
- Accessibility: motorway links for supplies and customers.

Unimportant factors

- Access to raw materials and power.
- Railway links.

Key words

quaternary industry
footloose

How and why the location of economic activity changes

Locally: in towns and cities

When urban areas were developing, streets where people lived grew up next to the factories. Few people had cars, and they walked to work. As towns expanded, large residential areas were built. There was no longer room for factories in the centre of cities so they relocated to the edges of the urban areas often in industrial estates.

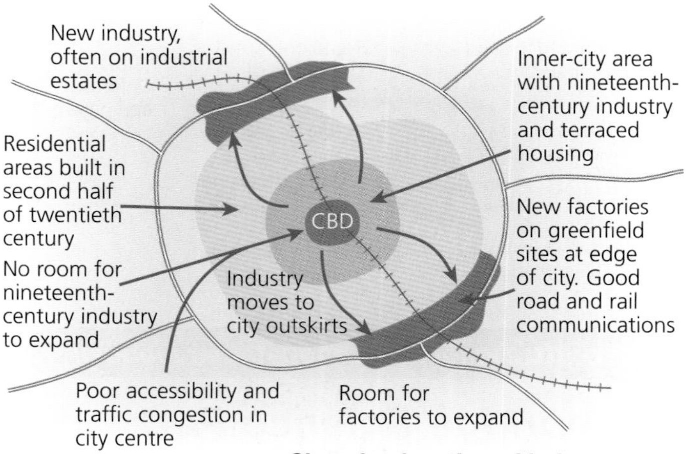
New industry, often on industrial estates
Inner-city area with nineteenth-century industry and terraced housing
Residential areas built in second half of twentieth century
CBD
New factories on greenfield sites at edge of city. Good road and rail communications
No room for nineteenth-century industry to expand
Industry moves to city outskirts
Poor accessibility and traffic congestion in city centre
Room for factories to expand

Changing location of industry

Nationally: changing location of iron and steelmaking in the UK

Iron and steel works 1967

- Located close to coast and shipbuilding industry, so close to markets
- Located close to coal and iron ore fields to reduce cost of transporting bulky raw materials
- Located close to iron ore fields and limestone to reduce cost of transporting bulky raw materials
- Located on coalfields to reduce cost of transporting bulky raw materials

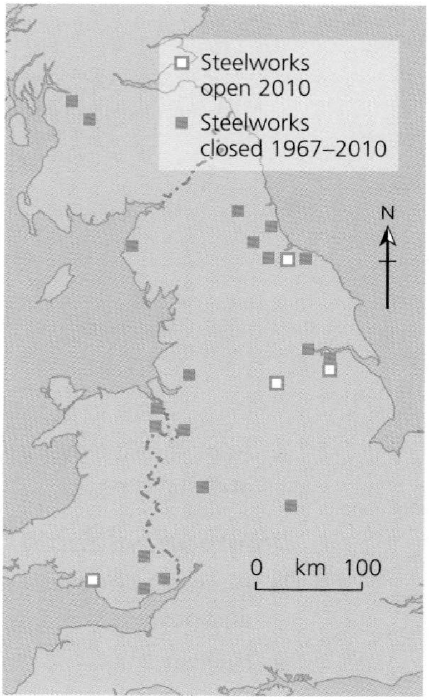
Steelworks open 2010
Steelworks closed 1967–2010
N
0 km 100

Major steel production plants 2010

- Competition from steel produced more cheaply in southeast Asia — inland works in the UK could not compete and they closed
- Local ore fields became exhausted so works dependent on imported ore
- Coal produced more cheaply overseas so plants on the coast continued importing coal and iron ore — other inland plants closed as transport costs increased
- Specialist plants in Sheffield and Rotherham were an exception — they used scrap metal to make high-grade specialist steels and forgings
- Only three modernised integrated steel works survive on the coast where imported raw materials can be assembled cheaply — one, Redcar, was mothballed in 2010 and its future, like that of the others, is in doubt

Internationally

Companies are always looking for the best location, particularly where costs are lowest. When the European Union expanded in 2004 the East European countries such as Poland, Slovakia, Estonia and the Czech Republic had lower tax rates and lower labour costs than other European countries such as the UK and Germany, and offered companies incentives such as free land and subsidies to help with costs.

Car manufacturing companies such as Volkswagen, Hyundai and Peugeot quickly realised the advantages and built plants in Slovakia. They were joined by firms such as Samsung making electrical goods such as televisions and printers, and Whirlpool making household goods such as washing machines and ovens.

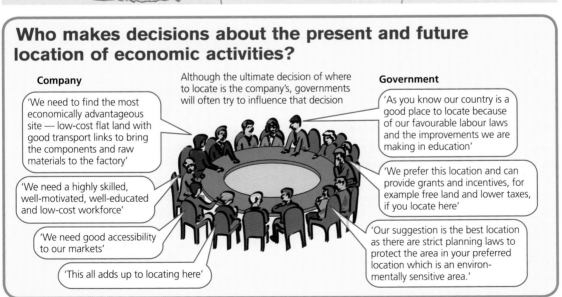

Factors attracting firms to Slovakia

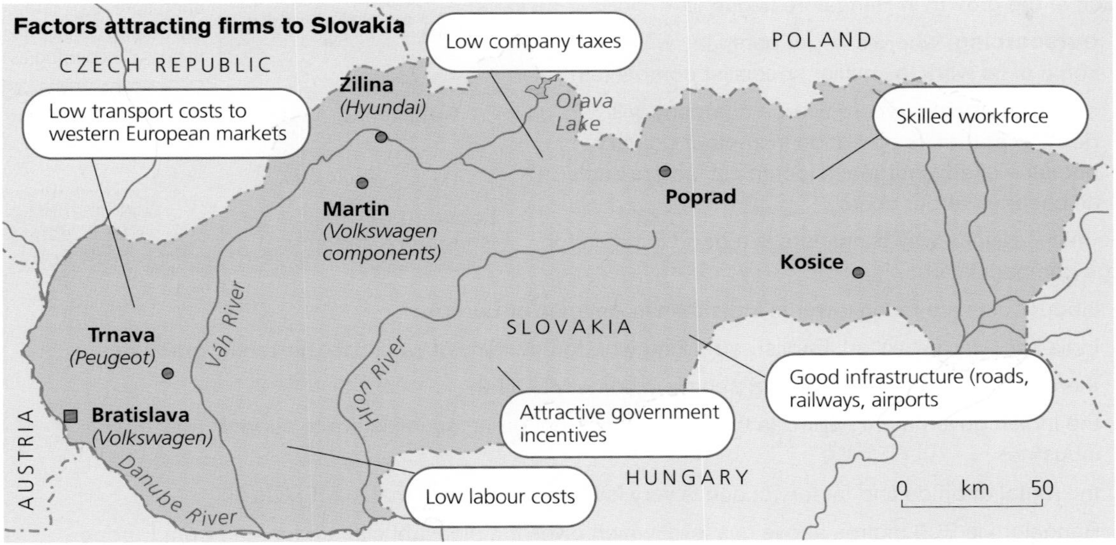

Who makes decisions about the present and future location of economic activities?

Although the ultimate decision of where to locate is the company's, governments will often try to influence that decision

Case study: *the location of an economic activity in an LEDC*

ICT industry in Bangalore, India

Fact file: India

- Population 1 billion
- 44% of population earn less than US$1 per day
- 70% of people are farmers

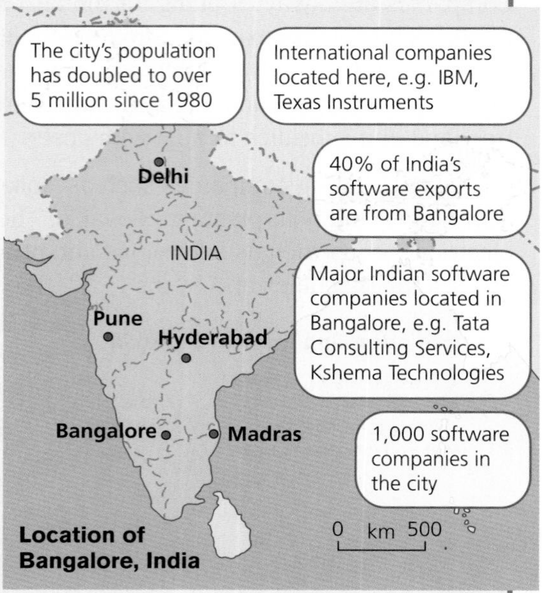

The city's population has doubled to over 5 million since 1980
International companies located here, e.g. IBM, Texas Instruments
40% of India's software exports are from Bangalore
Major Indian software companies located in Bangalore, e.g. Tata Consulting Services, Kshema Technologies
1,000 software companies in the city
Delhi
INDIA
Pune
Hyderabad
Bangalore
Madras
0 km 500

Location of Bangalore, India

Reasons for locating in Bangalore

Much of the growth in Bangalore is due to:

- **outsourcing**, where a larger company will contract some of its work to smaller specialist companies
- or to **offshoring**, where a large company will develop part of its operation in another country because of advantageous locational factors such as cheaper labour costs

The main reasons why Bangalore is a good location for large companies are:

- labour costs are much lower in India than in America or Europe
- India has a large, skilled, English-speaking workforce, many of whom are university graduates
- there is a shortage of ICT-skilled people in some MEDCs
- the Indian government supports the ICT industry, encouraging investors to support new ICT industries
- the rental of office and factory space is very low compared with that in MEDCs
- Bangalore is 900 metres above sea level which gives it a pleasant climate for staff from Europe and America
- as more ICT businesses grow in Bangalore, more workers are trained, knowledge can be exchanged and offshoot industries servicing the main companies can develop, so the advantages of **agglomeration** can be utilised

The benefits and issues of the ICT industry's growth in India

Benefits

- The creation of well-paid jobs in India
- Some of the most highly skilled workers have set up their own companies in India
- Bangalore has developed as a city with new buildings
- The software and ICT industries have contributed to the Indian economy and some are helping the government to train the people of India

Issues

- People in Europe and America lose jobs
- Some skilled workers go to work in the ICT industry abroad so their skills are lost to India
- Many people from the countryside have been attracted to Bangalore causing the city difficulties in servicing and housing the increase in population
- The benefits are largely restricted to the English-speaking middle-class group and in certain cities mainly in the south

Case study: *the location of an economic activity in an MEDC*

BMW cars near Birmingham

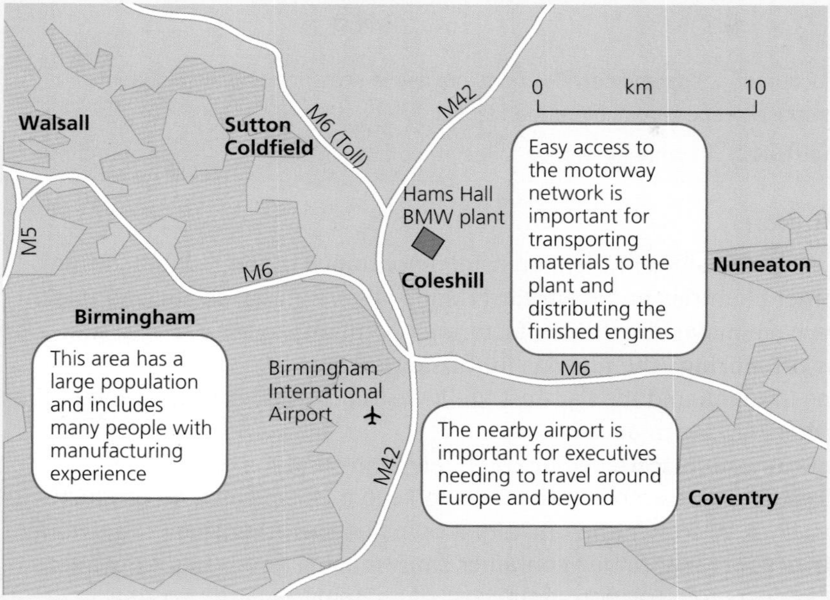

Walsall
Sutton Coldfield
M6 (Toll)
M42
0 km 10
Easy access to the motorway network is important for transporting materials to the plant and distributing the finished engines
Hams Hall BMW plant
Coleshill
M5
M6
Nuneaton
Birmingham
This area has a large population and includes many people with manufacturing experience
Birmingham International Airport
M6
The nearby airport is important for executives needing to travel around Europe and beyond
M42
Coventry

Location of the BMW (Hams Hall) plant, Coleshill, Birmingham

Fact file: UK

- Population 60 million
- Almost 90% of the people live in an urban area
- GNP US$24,430

Reasons for locating near Birmingham

The reasons why BMW located the factory at Hams Hall, Coleshill are:

- the site is close to good road, rail and air links, which means that the finished engines can be easily and quickly transported to factories around the world and key executive workers can move swiftly to factories in other countries
- Birmingham, with its long history of metal working, has a large population of skilled workers who can easily be trained in engine manufacture
- the factory was built on a **brownfield site** which was once a power station. With few planning restrictions, it offered BMW cheap flat land on which to build
- the government offered financial incentives if BMW located there, which made it cheaper and secured jobs for the local people

Benefits and issues of BMW locating near Birmingham

Benefits

- Secures well-paid jobs in the area for those who work in the factory, for those who work in the service sector where the workers will spend their wages and for small firms servicing the plant
- Helped to regenerate the area after traditional industries were in decline and provides revenue for the government

Issues

- This plant is just one of many BMW plants around the world and its future depends on decisions taken in Germany not in the UK

Key words

outsourcing
offshoring
agglomeration
brownfield site

Globalisation

Globalisation The production of goods and services on a worldwide scale to supply a global market leading to increasing interconnectedness and interdependence.

Multinational company (MNC) A company with factories and offices in several different countries.

Reasons for globalisation

Globalisation has taken place because of the increasing importance of the manufacturing industry in China and other countries in the Far East. Many large companies such as Hitachi, Toyota, Nike, Coca-Cola have moved their manufacturing operations to the Far East and China. This is known as **offshoring**. The reasons for offshoring are:

- **Lower wages**: wages in China and the Far East are lower, in fact about one-eighth of those in countries such as the USA, Japan and the UK.
- **Improvements in communications**: the internet, telephone and fax enable the headquarters of a company, for example in the USA, to keep a close watch on production, design, and research and development in a factory it owns on the other side of the world.
- **Improvements in transport**: enormous container ships moving goods across the seas have reduced transport costs to a fraction of the cost of an item they are carrying.
- **Free trade**: over the past 25 years trade barriers and tariffs between countries have fallen considerably, with the result that international trade has increased significantly.

Globalisation and the growth of multinational companies

Large companies look for ways to increase profits and lower costs to compete with their rivals in the global marketplace. This graph shows how they have developed their operations to take advantage of globalisation.

The development of MNCs

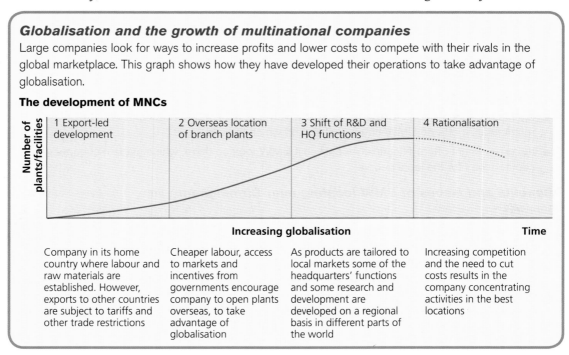

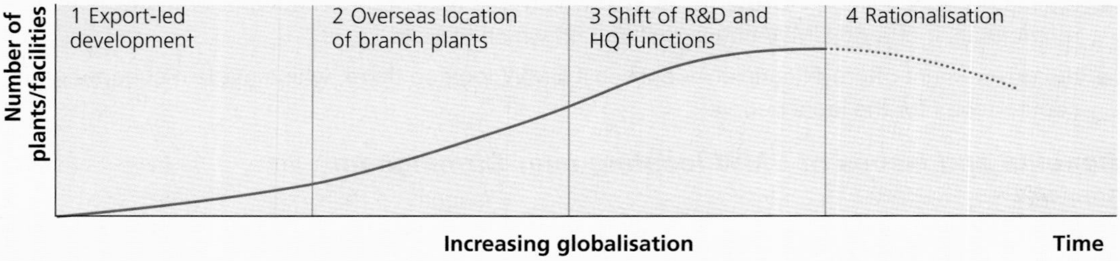

1	2	3	4
Company in its home country where labour and raw materials are established. However, exports to other countries are subject to tariffs and other trade restrictions	Cheaper labour, access to markets and incentives from governments encourage company to open plants overseas, to take advantage of globalisation	As products are tailored to local markets some of the headquarters' functions and some research and development are developed on a regional basis in different parts of the world	Increasing competition and the need to cut costs results in the company concentrating activities in the best locations

Case study: *globalisation*

The Apple iPod

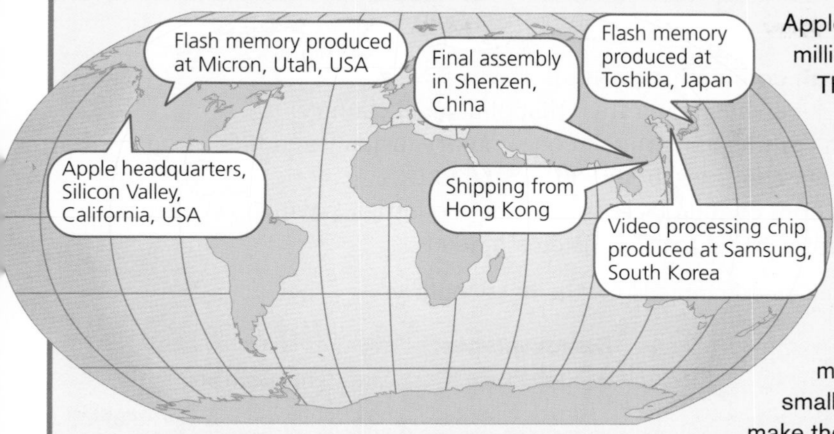

The worldwide locations used in producing an iPod

Apple had sold more than 240 million iPods by January 2010. The company has taken advantage of globalisation to build factories or use suppliers in countries with lower labour costs, good transport links, access to raw materials and a growing market.

Different companies manufacture and supply the small parts that are assembled to make the iPod, which was designed at the headquarters of the company.

Apple headquarters: where the decisions are made and the research, development and design of the iPod takes place.

Located in Silicon Valley, California, USA because:

- similar companies are located nearby and knowledge can be shared
- education establishments are nearby to offer training to future employees
- good climate, landscape and quality of life attract the best workers

Japan and Utah, USA: the memory chip accounts for around 50% of the cost of an iPod and is made by Toshiba in Japan and Micron in Utah.

These companies were chosen because:

- they are large well-established companies well used to making such complicated components
- they have a highly skilled workforce
- they are very hi-tech organisations with good access to well-developed technologies

South Korea: Samsung in South Korea makes the video processing chip. The company was chosen because it has the same hi-tech know-how and highly skilled workforce as Toshiba and Micron.

Shenzen, China: the iPod Nano is assembled in Shenzen, China. The components are assembled in one of the world's largest factories, employing over a quarter of a million employees.

They are assembled here because:

- labour costs are considerably lower than in Apple's home country, the USA
- there is a very skilled staff trained in the best methods of assembly
- nearby to the south is the huge port of Hong Kong to send the finished products to shops worldwide

Key words

globalisation
multinational company (MNC)
offshoring

Multinational companies

Multinational (or **transnational**) **companies (MNCs):**

- are very large, some with an income greater than that of a small country
- trade across the world, with their headquarters in one country (usually an MEDC) and branch factories in many other countries, both MEDCs and LEDCs
- can benefit in several ways from owning factories in other countries: paying low wages, avoiding some taxes and tariffs, being near or inside a market

There are disadvantages as well as advantages for the host country.

Advantages:

- provide jobs and good wages
- provide training to improve skills
- develop new roads and services and bring investment into the country
- increase foreign trade and bring foreign currency
- support other industries in the host country (the multiplier effect)

Disadvantages:

- often locate headquarters, research and development in home country but pay low wages for manufacturing in other countries
- bring foreign nationals to fill higher-paid jobs
- make goods for export, not for the host country
- take profits out of the host country
- close overseas factories first in difficult times

The possible futures for globalisation and the influence of MNCs

MNCs are increasing their market share in the global economy. For example, the increasing influence of large supermarket chains such as Tesco (second-largest retailer in the world for sales) and Wal-Mart (from the USA, trading as Asda in the UK, now the largest retailer in the world). These giants sell many types of goods from kitchenware (including cookers and refrigerators) to clothes at low prices in countries across the world. Small local high-street shops cannot compete and are forced to close, so changing the face of towns.

The global shift in manufacturing from MEDCs to NICs and LEDCs is set to continue and will affect the economy of MEDCs. For example, the manufacture of plastic toys, clothes and many electrical goods is now carried out in China, South Korea and other southeast Asian countries. Although headquarters' staff may remain in MEDCs the trend is for many of their jobs to move nearer the source of supply. MEDCs are now focusing on highly sophisticated manufacturing to prevent becoming too reliant on the tertiary sector.

Jobs are increasingly being filled by well-educated and highly motivated young people — a high percentage of the workers in call centres in India are university graduates. The less well educated with low levels of skills have found difficulty as their jobs in manufacturing have moved to other countries and higher levels of immigration have resulted in more competition for lower-waged jobs.

Key word

multinational company (MNC)

Case study: *MNC investment in a specific area and in an international context*

General Motors: a multinational company

Countries in which General Motors operates

Fact file

- General Motors is one of the top ten motor manufacturers in the world.
- General Motors' brands include Vauxhall, Chevrolet, Opel and Daewoo.
- The company headquarters are in Detroit, USA.
- General Motors operates in 32 countries worldwide.
- The company employs 266,000 people.

General motors in China

In a joint venture with a Chinese company General Motors built a new car-assembly plant to make 300,000 cars per year in Liuzhou in southeast China.

The advantages for General Motors are:

- the company can sell cars directly into the fast-growing Chinese market
- China's rapid economic development means that there is an increasing demand for commercial vehicles and, as wages increase, for domestic cars
- China has low wage costs so the cars can be made more cheaply, thereby increasing profits
- many electronic parts and steel are close by as they are already made in China

General Motors in the UK

General Motors operates two manufacturing plants in the UK, one at Luton and one at Ellesmere Port near Liverpool. The company had brought secure jobs, skills and wealth to these areas.

In 2009, because of a worldwide recession, the company had enormous debts.

General Motors was to be broken up and pieces of the European operation were to be sold off. Decisions about the future of the company were being taken in Detroit. The priority for the company was to save jobs in the USA.

Sales of new cars increased in the second half of 2009 and General Motors decided that the plant at Merseyside would remain open. Other plants in Europe would have staffing levels cut.

Conclusion

The experience of Ellesmere Port shows that MEDCs are as much at the mercy of multinational companies as are LEDCs, especially if the factory is a branch factory and the parent company is in a foreign country.

How different economic activities affect the physical environment

All economic activity affects the environment to some extent. Primary and secondary activities are generally considered to have the worst effect on the environment.

The environment may be affected through:

- **pollution**: for example, as discharges through chimneys (air pollution) or through pipes discharging chemicals into water courses or seas
- **damaging the landscape**: through quarrying by physically removing areas of rock or damaging ecosystems so that vegetation is removed or damaged, and wildlife habitats are destroyed
- **changing the environment**: expanding cities or building roads and retail parks which lay down huge areas of concrete and generate large volumes of polluting traffic

How different economic activities affect the environment

Primary industry	**Water pollution** from farming as pesticides and fertilisers are washed into water courses **Damage to the landscape** from quarrying as rock is removed, scarring the land, and from mining as waste is deposited on the surface **Changing the environment** from forestry as trees are removed, destroying an ecosystem, or overfishing leading to a depletion of fish stocks
Secondary industry	**Air and water pollution** from heavy industry from chimneys as part of the manufacturing process or from the burning of fossil fuels. In processing industries chemicals are discharged into water courses or lakes
Tertiary industry	**Changes to the landscape** from the retail industry as large supermarkets, retail outlets and their car parks spread concrete over the land **Air pollution** from retail outlets caused by exhaust fumes from customers' cars and delivery lorries and from the transport used to carry food and other goods to supermarkets from all over the world
Quaternary industry	**Pollution** — the effect is less obvious but components for computers have to be transported to assembly plants, electricity has to be created for them to use, which has an impact on the atmosphere, and their components contain harmful metals which are difficult to dispose of

While MEDCs can often begin to find sustainable solutions to some of these issues through research and development and government action, they always incur a cost. Many polluting activities have been relocated to, or have grown, in the low-wage economies of the LEDCs which are now major world polluters, e.g. India and China. These countries, which are striving to develop their economies, do not wish to, or cannot afford to, spend money on costly methods of reducing pollution.

Case study: *acid rain*

Economic activity affects the environment at an international scale

Acidic gases including sulphur dioxide and oxides of nitrogen are released by the burning of fossil fuels in industrial plants, power stations and vehicles. These gases may dissolve in water droplets in the atmosphere which later fall as acid rain.

An international issue: Europe

All countries in Europe are concerned about acid rain. A country may pollute its own atmosphere but the pattern of westerly winds means that the acid rain created in one country falls in another. For example, pollution from the UK affects forests and lakes in Sweden.

The causes of acid rain

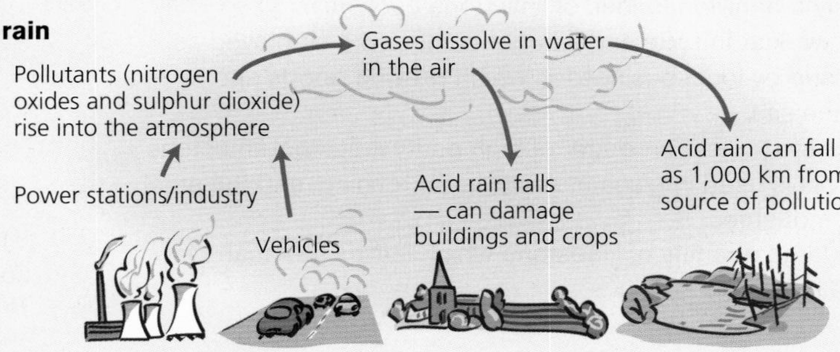
Pollutants (nitrogen oxides and sulphur dioxide) rise into the atmosphere
Gases dissolve in water in the air
Power stations/industry
Vehicles
Acid rain falls — can damage buildings and crops
Acid rain can fall as far as 1,000 km from the source of pollution

Effects of acid rain

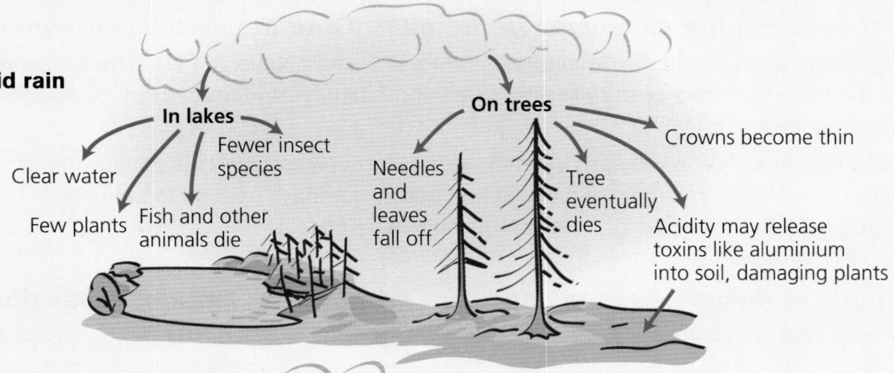
In lakes
Clear water
Few plants
Fish and other animals die
Fewer insect species
On trees
Needles and leaves fall off
Tree eventually dies
Crowns become thin
Acidity may release toxins like aluminium into soil, damaging plants

Reducing acid rain

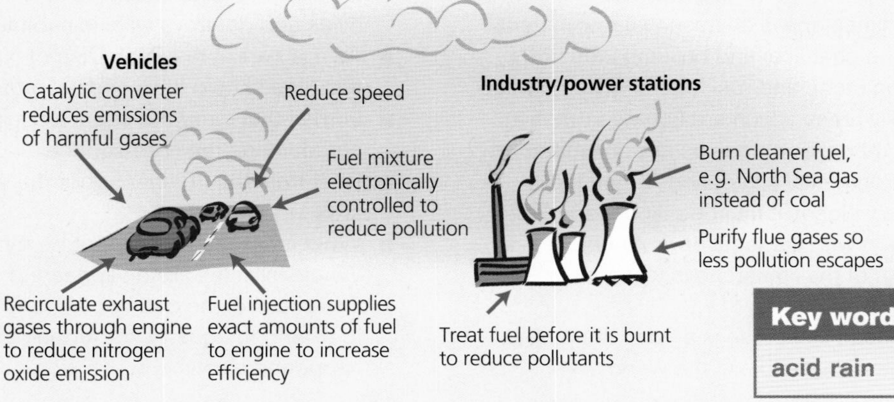
Vehicles
Catalytic converter reduces emissions of harmful gases
Reduce speed
Fuel mixture electronically controlled to reduce pollution
Recirculate exhaust gases through engine to reduce nitrogen oxide emission
Fuel injection supplies exact amounts of fuel to engine to increase efficiency
Industry/power stations
Burn cleaner fuel, e.g. North Sea gas instead of coal
Purify flue gases so less pollution escapes
Treat fuel before it is burnt to reduce pollutants

Key word

acid rain

Case study: *a specific development where conflict exists between economic development and environmental damage*

Tunstead Quarry, Derbyshire

Fact file

- Tunstead Quarry is located on the boundary of the Peak District National Park.
- The quarry is the largest limestone quarry in Europe.
- The carboniferous limestone at Tunstead is very pure, it is over 90% calcium carbonate.
- The quarry, together with the adjacent Old Moor Quarry, produces approximately 5.5 million tonnes of limestone each year.
- 400 people work at the site and a larger number are employed as contractors and by local businesses which provide goods and services to the site.
- The quarry is the largest producer of high purity industrial limestone in Europe. This is used in glass making, sugar refining, and the steel and cement industries.
- Much of the large quantity of limestone which leaves the quarry is carried by rail.

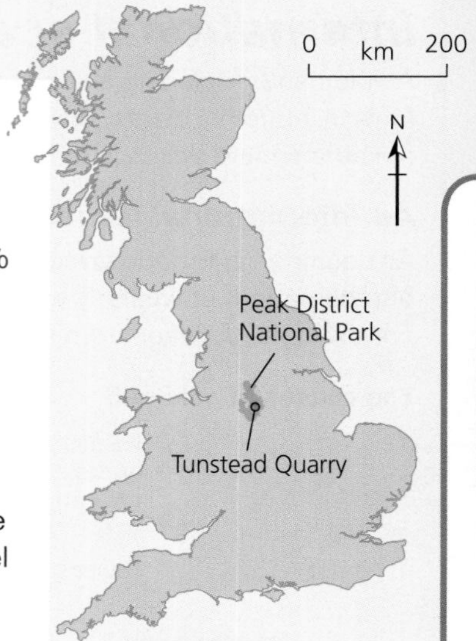

Location of Tunstead Quarry

Conflict between the expansion of the quarry and the local environment

The edge of the quarry is on the boundary of the Peak District National Park. The limestone areas of the Park, with their pretty villages, caves and beautiful landscapes, are protected by strict planning laws.

The quarry has steadily expanded to meet the increasing demand for its products and is running out of workable reserves of rock. So the quarry company applied to continue extraction across the valley, which is inside the National Park.

Arguments in favour of expansion

- Provides jobs in a rural area where work is scarce.
- The limestone is an essential raw material for the chemical and building industries.
- Good stone and rock can only be quarried in hilly areas which are usually attractive, and modern companies are committed to environmental protection. When the quarry closes it must be landscaped to tone in with the local environment.
- Much of the limestone is taken away by rail.
- Tunstead limestone is among the purest and most valuable in the world.

Arguments against expansion

- The ever-expanding quarry creates an eyesore in the landscape that is visible for miles and destroys wildlife habitats.
- As it is part of the Peak District National Park this area should have special protection.
- Lorries that carry the stone are large, noisy and damage the road surface.
- Dust from the crusher spoils the surrounding area.
- What is taken away cannot be replaced. Whole hillsides are disappearing.
- If successful the quarry company may seek to extend the quarry still further, as has happened elsewhere.

The future

Despite opposition from The Peak District National Park, conservation groups and environmentalists, the company was granted permission to expand the quarry. There is now a large quarry being worked within the National Park. Environmental groups are now worried that further expansion will be applied for and granted.

Reducing the impact

Facts

- There are currently 1,300 active quarries in the UK. More than 64,000 hectares of land has planning permission for quarries.
- Quarrying is a temporary land use, even though temporary may mean up to 100 years.
- A major question concerning environmentalists is what happens to the area when quarrying ends.
- Today permissions are only granted if there is a clear plan submitted by the company to sustainably restore the landscape in some way.

The quarry — caring for the landscape

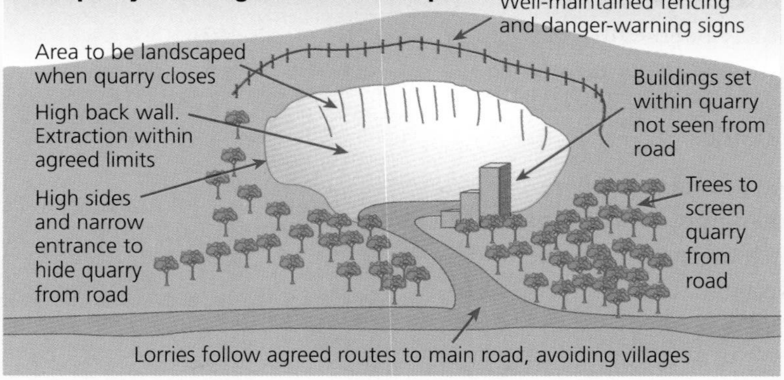
Well-maintained fencing and danger-warning signs
Area to be landscaped when quarry closes
High back wall. Extraction within agreed limits
Buildings set within quarry not seen from road
High sides and narrow entrance to hide quarry from road
Trees to screen quarry from road
Lorries follow agreed routes to main road, avoiding villages

GIS: Geographic Information System

GIS is a software package that allows storage, analysis and presentation of information about a location. The technology uses digital information from satellite and aerial imagery.

When a quarry site is being restored, for instance to a nature reserve, GIS enables the geology, soil type, contours and height of the water table to be overlaid on an aerial photograph. This would accurately inform decisions about the creation of the most suitable types of vegetation and habitat. The great biodomes of the Eden Project in Cornwall are probably the best known use of an old quarry, but examples of abandoned quarries becoming country parks, nature reserves and leisure lakes are common.

Key word

Geographic Information System (GIS)

Global climate change

The causes of global climate change

Global temperatures are rising. 1998 was the warmest year of the twentieth century and all but one of the warmest years on record have occurred since 1983. The world's leading scientists agree that this increase in global temperatures has largely been caused by the activities of people, especially in industrialised countries.

People's activities are adding **greenhouse gases**, such as carbon dioxide and methane, to the atmosphere.

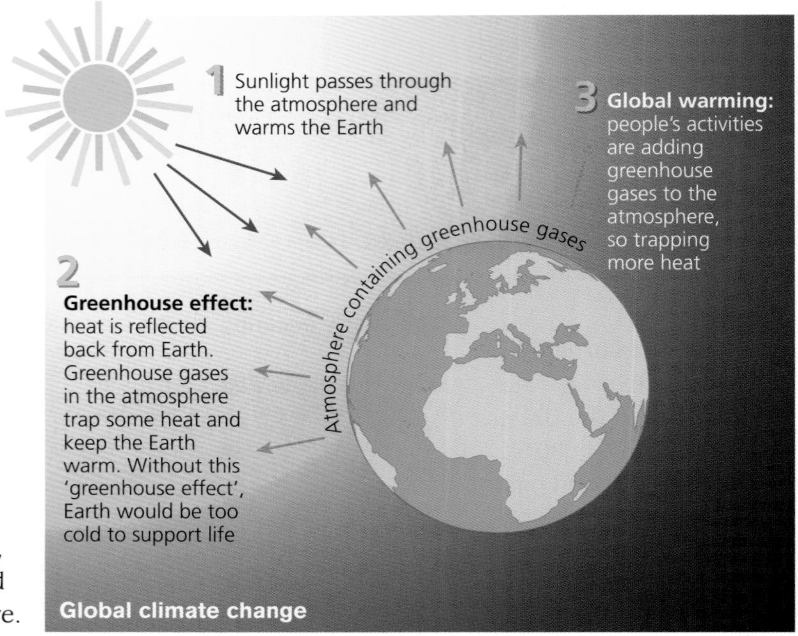

1 Sunlight passes through the atmosphere and warms the Earth
2 Greenhouse effect: heat is reflected back from Earth. Greenhouse gases in the atmosphere trap some heat and keep the Earth warm. Without this 'greenhouse effect', Earth would be too cold to support life
3 Global warming: people's activities are adding greenhouse gases to the atmosphere, so trapping more heat
Atmosphere containing greenhouse gases

Global climate change

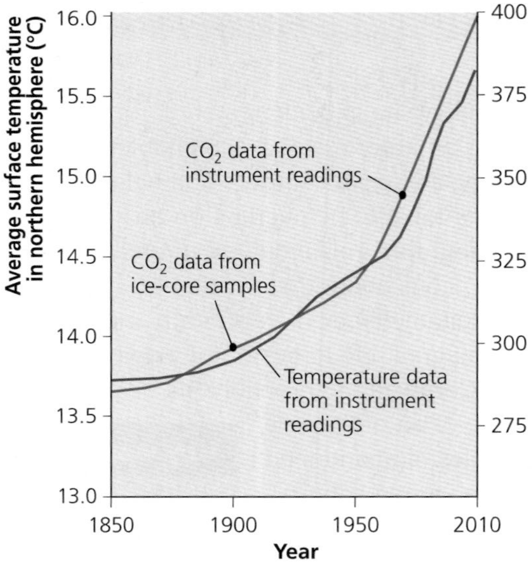

Average surface temperature in northern hemisphere (°C)
16.0
15.5
15.0
14.5
14.0
13.5
13.0
400
375
350
325
300
275
CO_2 data from instrument readings
CO_2 data from ice-core samples
Temperature data from instrument readings
1850
1900
1950
2010
Year

Carbon dioxide and climate change

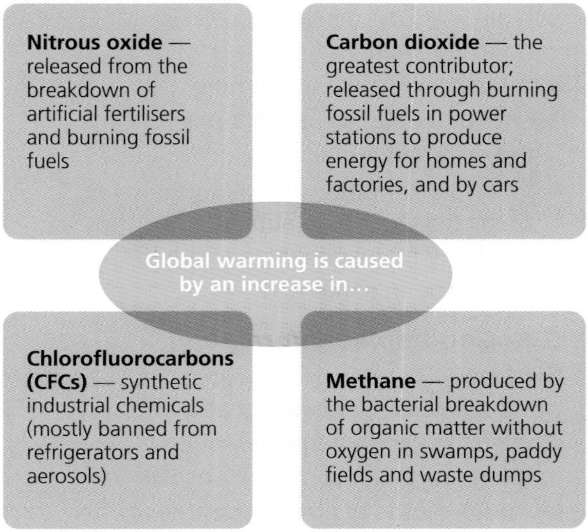

Nitrous oxide — released from the breakdown of artificial fertilisers and burning fossil fuels
Carbon dioxide — the greatest contributor; released through burning fossil fuels in power stations to produce energy for homes and factories, and by cars
Global warming is caused by an increase in…
Chlorofluorocarbons (CFCs) — synthetic industrial chemicals (mostly banned from refrigerators and aerosols)
Methane — produced by the bacterial breakdown of organic matter without oxygen in swamps, paddy fields and waste dumps

Greenhouse gases

Scientists believe that the increase in carbon levels released into the atmosphere is so closely mirrored by the increase in global temperatures that the link is indisputable, and that global warming is not part of a natural cycle as some people claim.

Who is to blame? We all are but...

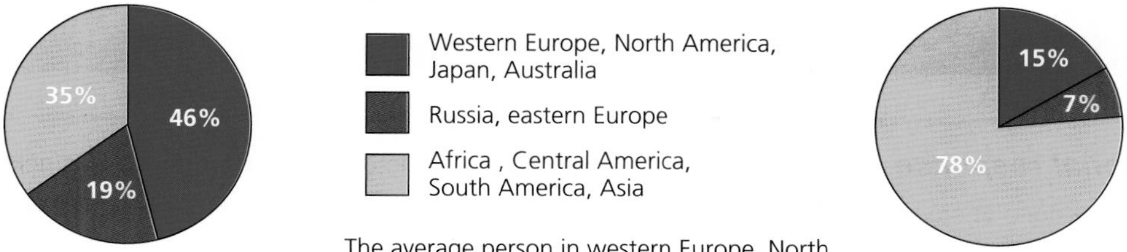

Responsibility for global warming based on energy consumption

The average person in western Europe, North America, Japan and Australia uses 10 times more energy than the average person in Africa, Central America, South America and Asia

Share of world population

The effects of global climate change

Increasing global temperatures are beginning to have effects all over the world.

Effects of global warming

Global effects 2: wildlife

Many species of plants and animals will find it hard to adapt to changes in climate and habitat. Polar bears, which hunt seals on the sea ice, are already finding it difficult to adapt to the thawing of the ice in the Arctic.

Global effects 1: climatic belts

The climatic belts will expand north and south. The expansion of the drought-prone areas to the north and south of the desert regions could affect water supplies and agriculture over areas of the southwestern USA, the Mediterranean, southern Australia, southern Africa and parts of South America.

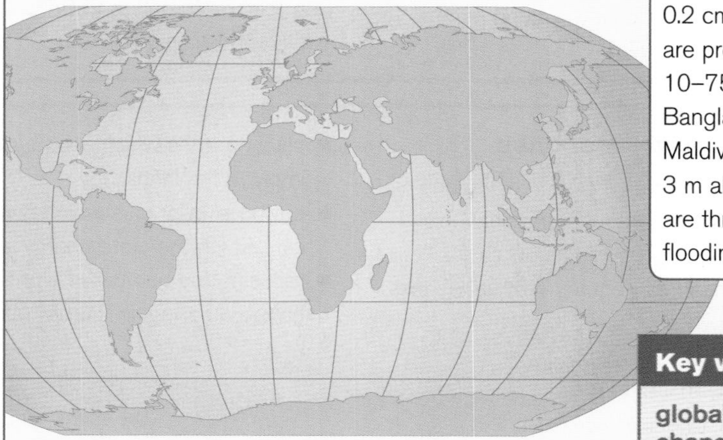

Global effects 3: sea levels

Sea levels are rising by 0.2 cm per year and are predicted to rise 10–75 cm by 2100. Bangladesh and the Maldives (highest point: 3 m above sea level) are threatened by flooding.

Global effects 4: patterns of rainfall

Changes to the amount and distribution of rainfall. Areas with plenty of rainfall will receive more and regions with low rainfall will receive less, affecting the type and yield of the crops that can be grown.

Key words

- global climate change
- greenhouse effect
- greenhouse gases
- nitrous oxide
- carbon dioxide
- chlorofluorocarbons (CFCs)
- methane

Responses to global climate change

International response

Nothing can be done to stop global warming, but action can be taken to slow it down in the future. This relies on a joint approach by all the large countries in the world, which is difficult to achieve.

- The Kyoto Agreement, signed by many countries in 1997, was an important step towards reducing emissions of greenhouse gases and most European countries are now taking action. By 2005, 141 countries had ratified the agreement, representing over 61% of global emissions. India and China who (with their rapidly growing economies) are major polluters, were exempt. The USA signed the Kyoto Agreement but was reluctant to make any cuts, despite the fact that it produces at least 25% of the main polluting gases. It is worried that cuts will damage its economy.
- In 2009 an international conference was held in Copenhagen to strengthen the international agreement reached at Kyoto and to include India, China and the USA. No legally binding agreement was reached. Countries pledged to reduce emissions and limit global warming to 2°C. Some LEDCs wanted aid to help reduce emissions and others such as India and China were reluctant to commit to something they felt would only slow their fast-growing economies.
- **Carbon trading** is a system whereby governments set a limit to the amount of carbon a company can emit. If this is exceeded then the company must buy 'credits' from those who pollute less. If the company emits less carbon then it can sell the 'credits' saved. The value of carbon traded in 2008 worldwide was US$60 billion.

Local and individual response

Individuals can help to reduce carbon emissions by:

- using public transport wherever possible — one bus can carry as many people as 30 cars
- making our homes more energy efficient so less power is used
- planting trees which absorb carbon dioxide and improve the landscape
- buying smaller cars which burn less petrol and cars which are more energy efficient
- reducing the number of flights we take, such as having a holiday in the UK rather than abroad

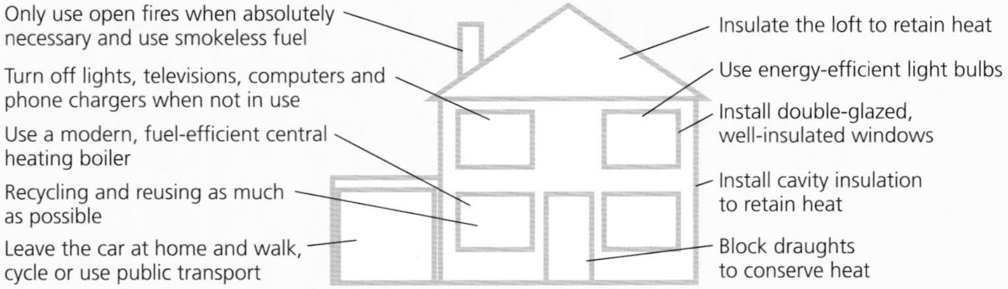
Only use open fires when absolutely necessary and use smokeless fuel
Turn off lights, televisions, computers and phone chargers when not in use
Use a modern, fuel-efficient central heating boiler
Recycling and reusing as much as possible
Leave the car at home and walk, cycle or use public transport
Insulate the loft to retain heat
Use energy-efficient light bulbs
Install double-glazed, well-insulated windows
Install cavity insulation to retain heat
Block draughts to conserve heat

Reducing carbon emissions at home

National response

The British government has agreed to cut emissions of greenhouse gases, particularly carbon dioxide, by:

- setting targets: the British Government is committed to reduce emissions by 60% by 2050, and is committed to producing 20% of its electricity from renewable sources by 2020
- using less coal and oil to generate electricity and using natural gas or renewable energy instead. In 2009 the government made a commitment to building new nuclear power stations which generate little greenhouse gases but are very expensive, and there remain problems with disposing of spent fuel
- using new technology such as **carbon capture** in which the carbon emissions from power stations are captured and stored underground in old oil and gas fields
- improving insulation in public buildings such as schools and hospitals
- persuading people to save energy by insulating their houses, turning lights out and only boiling as much water in the kettle as they need

Key words

carbon trading
carbon capture

***Case study:** the effect of global climate change*

How will global climate change affect Britain?

- Rising sea levels of 20 to 40 centimetres mean low-lying areas will be flooded, for example the Fens and parts of the Thames Estuary.
- There will be more gales and storms, particularly in the winter. The winter of 2000/2001 was the wettest since records began.
- Summers will be warmer and drier. Average temperatures may rise by 1.6°C by 2050. Farmers in the south will need to irrigate their crops. Water companies will have to store more water to meet increased demand.
- Britain could become a more popular holiday destination, pavement cafés could flourish and air conditioning will be needed in the south.
- There will be more rain and more likelihood of flooding in the autumn and winter. More money will have to be spent on flood protection, such as building embankments.
- Coniferous trees will grow more quickly, so up to 15% more wood will be produced, but deciduous trees could die from drought and disease.
- Vines, sunflowers and maize could be grown in the south.
- Insect pests could flourish.

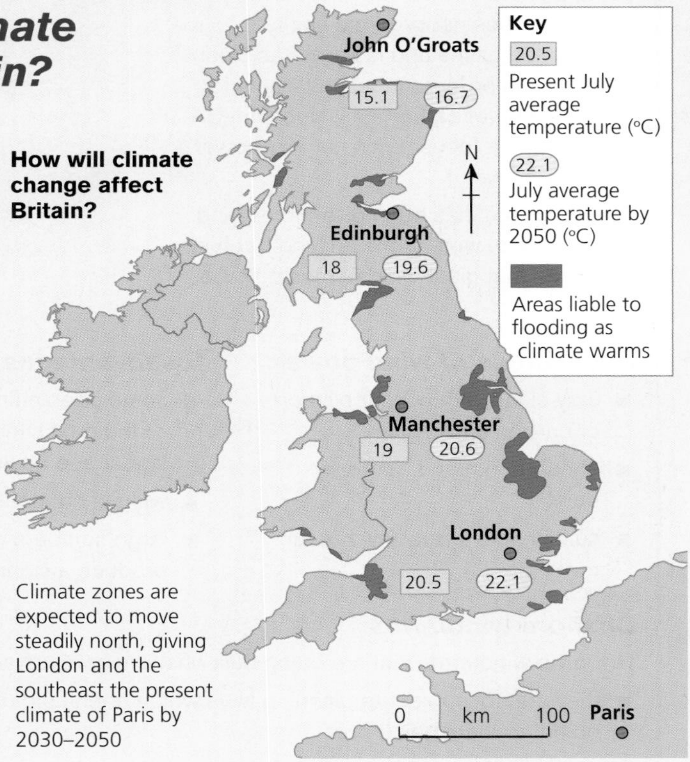

Case study: *a response to global climate change — wind farms*

Wind farms in the UK

One of the major contributors to carbon emissions and therefore global climate change is the generation of power. Over 60% of the electricity we use is produced by burning vast quantities of fossil fuels such as coal, oil and gas. Another 35% is generated by nuclear power and only approximately 5% is produced from renewable sources such as wind power, hydroelectric power, solar power and tidal power.

Using renewable sources to create electricity does not produce any greenhouse gases and is known as clean electricity. In order to meet its targets for the reduction of carbon emissions the UK government is focusing on nuclear power and wind power.

The 'go ahead' has been given to expand the number of wind farms on land but it is in the sea that giant wind farms are to be built.

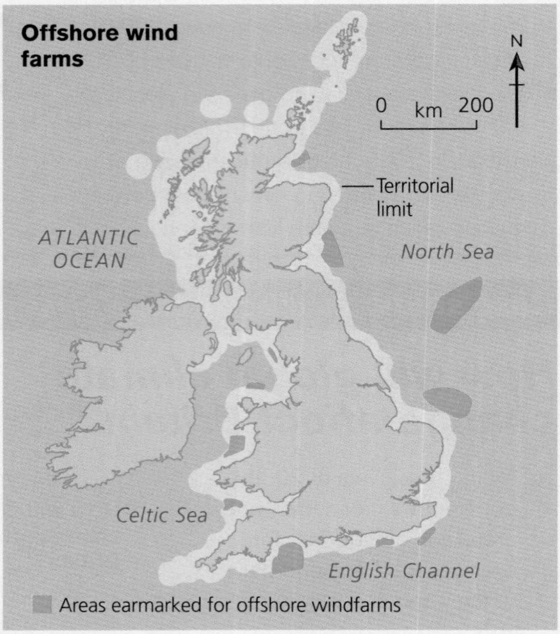

Advantages of wind power:

- very clean with no air pollution or carbon emissions
- cheap to run and generally quiet
- sustainable — fuel will not run out

Disadvantages of windpower:

- some people find the turbines visually polluting — the best sites are often upland areas of high landscape quality that are already protected
- wind is not constant and may stop blowing
- large numbers of turbines are needed to produce a significant amount of power

Offshore windfarms

The large windfarms that are being built offshore have advantages over those on land:

- they are not subject to planning laws which often hold up developments on land when people object to them being built
- the wind farms can cover large areas of sea without costly land to buy
- the individual turbines can be very large without being visually polluting

There are some perceived disadvantages. They may have an impact on:

- wildlife such as birds migrating or moving to feeding grounds
- the capability of radar to detect incoming ships or low-flying aircraft

Test yourself

1 **Give four indicators of development. What is meant by sustainable development?**

2 **What is the difference between 'bilateral aid' and 'multilateral aid'?**

3 **How does the employment structure in most LEDCs differ from that in MEDCs?**

4 **What are the differences between the 'formal sector' and the 'informal sector' of industry in an LEDC?**

5 **Give two examples of each of the following: primary industry, secondary industry, tertiary industry and quaternary industry.**

6 **Draw a star diagram to show as many reasons for industrial location as you can.**

7 **Give one example of how the location of industry has changed in the last 100 years.**

8 **Give one example of globalisation. What is a multinational company? Give four advantages of a multinational company locating in a country.**

9 **Give three reasons for granting an extension to a quarry in an area of natural beauty.**

10 **Explain the main reason for climate change. Give two sustainable responses to climate change.**

Examination question

(a) **Four types of industry are primary, secondary, tertiary and quaternary. Write the correct type alongside each statement below:**

(i) **Working in offices and shops serving the public.**
(ii) **Growing and cutting trees for making furniture.**
(iii) **Carrying out research into new medicines.**
(iv) **Making electrical goods for use in the home.** *(2 marks)*

(b) **Look at the diagram which shows the types of industry in Peru, an LEDC.**

(i) **How does the diagram of types of industry in Peru differ from that of an MEDC?** *(2 marks)*
(ii) **Give three advantages of attracting transnational companies to open branch factories in an LEDC.** *(3 marks)*

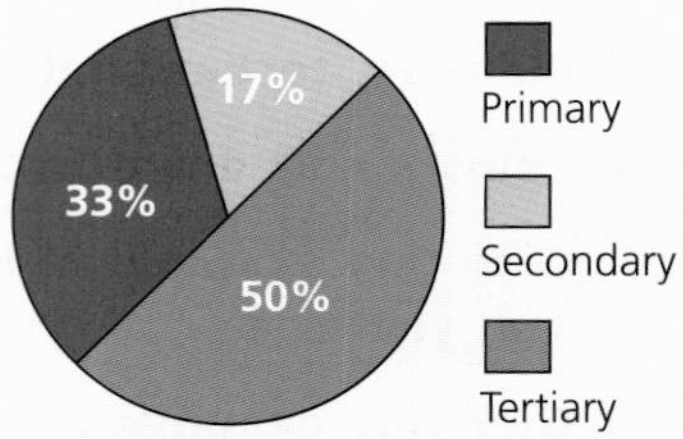

Types of industry in Peru

Case study: *location of industry*

Foundation tier:

(c) **Look at the map which shows the location of Cambridge Science Park.**

(i) **Suggest the advantages the Science Park brings to Cambridge.**
(ii) **Suggest reasons why the Science Park was located on the outskirts of Cambridge.** *(8 marks)*

Higher tier:

(c) **For a newly built factory you have studied:**

(i) **Name the factory or site.**
(ii) **Describe and explain why the factory opened at that location, and what benefits it brought to the area.** *(8 marks)*

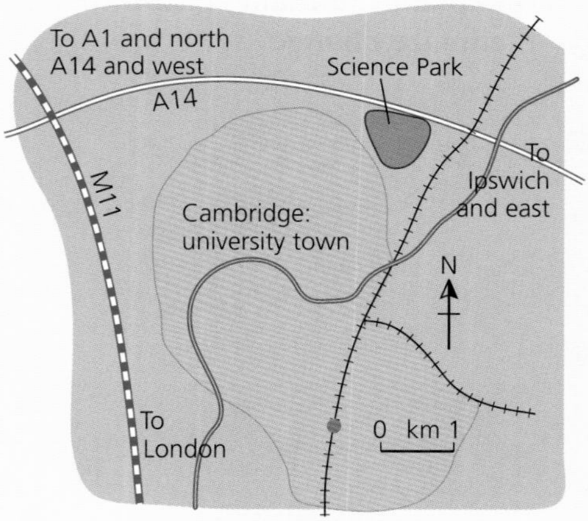

Location of Cambridge Science Park

Exam tip

In the higher tier question, notice that you need to name an actual example. There are two parts to the last sentence: 'explain why…' and 'what benefits…'. If you miss one, you will lose marks.

Key word index

Notes